Sensors and Measurement Systems

Second Edition

RIVER PUBLISHERS SERIES IN ELECTRONIC MATERIALS AND DEVICES

Series Editors:

Edoardo Charbon
EPFL, Switzerland

Mikael Östling
KTH Stockholm, Sweden

Albert Wang
University of California, Riverside, USA

Indexing: All books published in this series are submitted to the Web of Science Book Citation Index (BkCI), to CrossRef and to Google Scholar.

The "River Publishers Series in Electronic Materials and Devices" is a series of comprehensive academic and professional books which focus on the theory and applications of advanced electronic materials and devices. The series focuses on topics ranging from the theory, modeling, devices, performance and reliability of electron and ion integrated circuit devices and interconnects, insulators, metals, organic materials, micro-plasmas, semiconductors, quantum-effect structures, vacuum devices, and emerging materials. Applications of devices in biomedical electronics, computation, communications, displays, MEMS, imaging, micro-actuators, nanoelectronics, optoelectronics, photovoltaics, power ICs and micro-sensors are also covered.

Books published in the series include research monographs, edited volumes, handbooks and textbooks. The books provide professionals, researchers, educators, and advanced students in the field with an invaluable insight into the latest research and developments.

Topics covered in the series include, but are by no means restricted to the following:

- Integrated circuit devices
- Interconnects
- Insulators
- Organic materials
- Semiconductors
- Quantum-effect structures
- Vacuum devices
- Biomedical electronics
- Displays and imaging
- MEMS
- Sensors and actuators
- Nanoelectronics
- Optoelectronics
- Photovoltaics
- Power ICs

For a list of other books in this series, visit www.riverpublishers.com

Sensors and Measurement Systems
Second Edition

Walter Lang

University of Bremen
Germany

River Publishers

Published, sold and distributed by:
River Publishers
Alsbjergvej 10
9260 Gistrup
Denmark

River Publishers
Lange Geer 44
2611 PW Delft
The Netherlands

Tel.: +45369953197
www.riverpublishers.com

ISBN: 978-87-7022-607-3 (Hardback)
      978-87-7022-606-6 (Ebook)

©2021 River Publishers

Contents

Acknowledgment

Many people have helped me writing this book. First of all, I thank my wife Christine for encouragement and support. I thank all members of the IMSAS (Institute for micro sensors, –actors and –systems) for discussing with me, reading the text, and for contributing figures. I want to mention Andreas Schander, who read the whole text, and Eva-Maria Meyer, who took SEM images. Lisa Reichel helped me with the English language.

Figures were contributed by Gerrit Dumstorff, Daniel Gräbner, Christian Habben, Nico Hartgenbusch, Thomas Hertzberg, Martina Hübner, Reiner Jedermann, Frieder Lucklum, Eva-Maria Meyer, Anmona Shabnam Pranti, Andreas Schander, Timo Schary, Christoph Sosna, Jörg Stürmann, Rico Tiedemann, and Nayyer Abbas Zaidi.

Friendly colleagues helped me by reading chapters of the text: Klaus Knuth, Klaus Pawelzik, Michael Ortner, Svenja Willing, Jim Deak.

A number of students from the University of Bremen helped and provided examples from their work, including Jennifer Hinz, Wiebke Gehlken, Marcel Tintelott, Paul Meilahn, and Muhammad Uzair Talal Chishti.

Figures were provided by:
Wolfgang Wiedemann, Sensortechnik Wiedemann GmbH
Franz Laermer, R. Bosch GmbH
Martin Trächtler, HSG-IMIT, Villingen-Schwenningen
Jörg Schieferdecker, Heimann Sensor GmbH
Andreas Meile, Sensirion AG
Ralf Bergmann, BIAS, Bremen
Ursula Dicke, University of Bremen
Tim de Rijk, University of Bremen

List of Figures

List of Tables

List of Abbreviations

ABS	Anti-Blocking System
AC	Alternating Current
ADC	Analog to Digital Conversion
AR	Aspect Ratio
CAD	Computer Aided Design
CCD	Charge Coupled Device
CVD	Chemical Vapor Deposition
DC	Direct Current
DOS	Density of States
DRIE	Deep Reactive Ion Etching
ESC	Electronic Stability Control
FPGA	Field Programmable Gate Array
GMR	Giant Magnetoresistance
GPS	Global Positioning System
IR	Infrared Radiation
KOH	Potassium Hydroxide
LPCVD	Low Pressure Chemical Vapor Deposition
MCU	Micro Controller Unit
MEMS	Micro Electro-Mechanical System
NDIR	Non Dispersive Infrared Sensor
NEP	Noise Equivalent Power
NTC	Negative Temperature Coefficient Material
OIS	Optical Image Stabilization
Op-amp	Operational Amplifier
PDMS	Polydimethylsiloxane
PECVD	Plasma Enhanced Chemical Vapor Deposition
PI	Polyimide
Poly	Polycrystalline Silicon
PTC	Positive Temperature Coefficient Material
PVD	Physical Vapor Deposition
RFID	Radio Frequency Identification

RIE	Reactive Ion Etching
RMS	Root Mean Square
TCR	Temperature Coefficient of Resistivity
TMR	Tunneling Magnetoresistance
TTV	Total Thickness Variation

Introduction: The Idea of This Book

I was just writing a sentence "Sensors are among the most important devices of...", when the idea crosses my mind that I might look, and I took a paper and a pencil, went through the house, and made a tally sheet of sensors. We are certainly not high-tech addicts, but yet it is more than 90, including more than 40 temperature sensors, 3 fever thermometers, about 10 electronic watches and timers that also show room temperature, and a water heater, a dishwasher, and a stove with one sensor for each boilerplate in the kitchen. The gas stove for heating has three thermal sensors. Each PC, laptop, tablet, or smartphone has one or more security gadgets against overheating (hopefully). There are nine fire detectors and four motion detectors for light in the garden. The second group, astonishingly, includes 13 CCD cameras: tablets and smartphones have 2 each; there are some old telephones with one camera in some drawers, and three real cameras to take pictures. The same devices hold about 10 accelerometers. I found five pressure sensors: three in the gas stove, one in the dishwasher, and a blood pressure monitor. There are two force sensors: one in the kitchen scale and the other in the bathroom scale. Finally, there are about 10 Hall sensors controlling disk drives in the PC and CD players. And I did not even look at our car. We use sensors throughout. For an engineer, understanding them is a must, and it still may be highly interesting for others, too.

What does this book aim at, and for whom has it been written? Sensors and measurement systems is a course in the first semester of the master study of electrical engineering at the University of Bremen. The audience is international students, having a bachelor degree in electrical engineering and focusing now on a master curriculum in control, mechatronics, microelectronics, or microtechnology. All these fields develop rapidly, seeing an increase of computing and digital control. However, digital control not only means microelectronics and communication, but also sensing, because control needs information about what is going on. Thus, a major topic for electrical engineers is to understand sensors: what can they measure, how is it done, where are the limits, and how are they made? This book aims to answer these questions. But I do not address only electric engineers. People from physics, chemistry, medicine, biology, informatics, mechanical engineering…They all use sensors and may greatly benefit from understanding them. The field is too large to cover everything, so my aim is not completeness, but I want you to understand the principles of sensors. I want to enable you to communicate with the sensors community and find answers in case of sensor questions.

After reading this book, you should be able to:
Name and explain important thermal, mechanical, and magnetic sensors.
Work with characterization parameters for sensors.
Choose sensors for a given application and apply them.
Analyze existent sensor systems.
Understand micromachining technologies for sensors.

The number of sensors is large, we have to select and to find a way to structure the material. My approach is to assemble sensor families. What might the "family of thermal sensors" be? Thermal sensors measure temperature, of course, but they do much more. Several physical effects involve some energy transport, which causes some temperature change, and thus can be measured thermally. Flow, all sorts of radiation, electric power, chemical reaction, and other quantities can indirectly be measured by a temperature difference they generate.

My favorite example is the thermopile for the measurement of infrared radiation. It is used for non-contact temperature measurement, such as the ear fever thermometer. The input is a small amount of radiative energy; the output is a proportional voltage. We construct a thermopile as shown in Figure 1. This is a thin membrane, which is heated up slightly by the radiation. Then, we integrate some thermocouples on this membrane to produce a voltage.

The members of the thermal sensor family may be very different, but there are always three essentially major components arranged in a row, as follows:

- Power input: Whatever makes a power signal can be measured thermally. This may be radiation, electric current, and many others.
- Transduction structure: A small thermally insulated spot, which makes a large temperature difference from a small input power.
- Readout effect: A measurement effect, which translates the temperature difference in an electric signal.

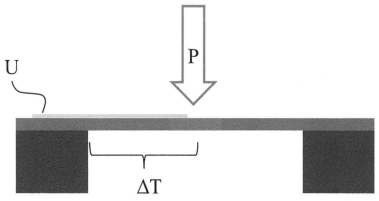

Figure 1   Thermocouple as an example of the family of thermal sensors. A small power input P generates a voltage output U.

The following are some other members of this family:

- Flow: thermal anemometer
- Electrical power: true RMS power meter
- Chemical composition of combustible gases: catalytic gas sensor
- Humidity: dew point detector
- Pressure: thermal vacuum sensor
- Inclination: thermal inclinometer
- Analysis of chemical processes: micro calorimeter

The example of the thermopile for measurement of IR radiation indicates how we could approach a measurement problem and find a suitable sensor solution. Figure 2 shows my general approach, which will also accompany us throughout the book.

First, we have to understand the measurement task. What exactly has to be measured? What specifications are needed to do the task in terms of sensitivity, span, signal-to-noise ratio, and working temperature? When we know this, we proceed in two columns: theory and manufacturing.

We have to understand the physics of the measurement problem and planned sensor quantitatively, which will allow us to write a mathematical model of the sensor response. Now we can derive the sensitivity in terms of voltage output per sensor input. For the thermopile, this would be V/W: Thermoelectric voltage by radiation power. At this point, can we already be happy with our understanding? Not yet, since we do not know how to distinguish a good sensor from a bad one. Sensitivity can be easily boosted

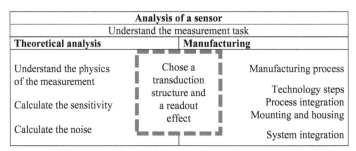

Figure 2   The general approach of this book to analyze a sensor.

by a transistor, but the sensor does not become any better. We need a criterion for quality, and this is noise: only high sensitivity combined with low noise makes a good sensor. Therefore, we need to complete our analysis by constructing a model for the noise and calculating the signal-to-noise ratio.

However, we have not yet finally decided about the structures. Theory explains how a structure might perform; technology tells us how it can be realized. We need to decide about both a transduction structure and a sensor effect, which is shown by the bold dashed field in Figure 2.

Now, let us focus on manufacturing. We start by drawing the sensor in CAD, which is called sensor layout. Manufacturing itself has three important aspects. Most micro sensors are made by silicon micromachining, so we need to understand all the processes done in a clean room with silicon wafers. However, many processes alone do not make a process flow. We may know how to make an aluminum thin film and how to manufacture high-temperature silicon nitride for insulation, but we cannot combine, because the aluminum would melt during the nitride deposition process. Thus, we have to understand the cooperation of subsequent processes and the full process integration.

When a wafer is finished, it leaves the clean room and is sawn into pieces to have hundreds of small sensors. Here, the second part of sensor making starts, which involves mounting and housing. Finally, a sensor also needs electronics and it must be integrated in the surroundings of a measurement system.

Knowing this, the organization of the book becomes apparent: The first field of application is the family of <u>thermal sensors</u>. How can we measure temperature in micro systems and why do we use a membrane for transduction? When we know what we want, it is time to move to Chapter 2, <u>Technology</u>. Do not be afraid, I do not intend to make you a silicon technologist within one book chapter. My aim in technology is to make you understand the possibilities, you should learn technology deep enough to be able to address the central part in Figure 2: selection of structures and effects. Furthermore, I want you to be able to speak to technology people and to understand their specific problems. Our paradigm example for Chapters 1 and 2 will be the thermopile sensor for IR radiation. After discussing processes (lateral approach), we will discuss full process integration (vertical approach) for this example. This will also include process accompanying metrology. Then, we will discuss possible problems: when we drive a process against the wall how does it happen?

Following this, we look at the family of mechanical sensors, which essentially measure force. There are two branches: (1) force by contact: <u>force sensors and pressure sensors</u> and (2) sensors for <u>inertial forces</u>: Newton force for accelerometers and Coriolis force for rotation sensors, also called micro gyroscopes, which measure the angular rate. Again, we will discuss the physics of a gyroscope to understand the sensor response and to estimate the noise equivalent angular rate. From this understanding, we can derive appropriate sensor structures, find a layout, and discuss the technology used to realize micro gyroscopes. The third family is sensors for the <u>magnetic field</u>.

The last chapter on <u>flow sensors</u> seems to dance out of line. It is kind of transversal with respect to the logic of the book. The idea is a recapitulation of the sensor families, rethinking effects and formulas, and using them in a new context. In fact, each family has flow sensors among its members: Thermal flow sensors allow us to revise the thermodynamics of thin membranes. Venturi tubes are an interesting application of pressure sensors. Coriolis flowmeters repeat the principles of gyroscopes. And even magnetic sensing is active in flow: the electromagnetic flow sensor is just a Hall sensor turned upside down.

Focusing on really important principles is a good didactic strategy, but you also want to see examples and you want to know about the hot topics discussed today. For this reasons I add some excursions at the end of the chapters. In these I show current trends, and I give some examples from my own groups ongoing projects.

1

Thermal Sensors

1.1 Measuring the Temperature

Before measuring temperature, we should ask what it is and what dimension it has. Thermodynamics shows that every particle is incessantly moving, and no atom or molecule is ever at rest. At low temperature, the movement is slow, and at high temperature, it becomes fast. Zero temperature would be no movement, but this cannot be the case in real world. "Temperature is just another measure for the average kinetic energy of the molecules"[1]. In principle, we would not need a specific dimension; we could just give the average energy of a molecule in the dimension of energy, which is Joule. Water boils at 1.54×10^{-20} J. When this was understood in the 19th century, it was already common to measure temperature in degrees, and physicists decided to remain with using degrees, which is more illustrative. Therefore, we need a constant between Joule and degree, which is the Boltzmann constant $k_B = 1.381 \times 10^{-23}$ J/K. Now, we can write the average kinetic energy of a particle per degree of freedom as:

$$\overline{W} = \frac{1}{2}m\overline{v^2} = \frac{1}{2}k_B T \tag{1.1}$$

A water molecule has 6 degrees of freedom (3 translations and 3 rotations), and we find that it melts at 273 K. When we start the scale at the absolute zero temperature, it is the Kelvin scale, or when we start at the melting point of water, it is the Celsius scale.

For household applications, consumer products, or automotive products, a typical sensor range would be -40 to $+120°$C. In industrial process control, higher temperatures have to be measured, up to 800°C. If temperatures become even higher, we normally will not mount a temperature sensor but measure temperature contactless via infrared radiation.

---

[1]C. Gerthsen, H. Vogel: Physik; Springer ed.

Classical methods of thermometry use the thermal expansion of materials. In liquid thermometers, we put a little fluid in a glass capillary. When temperature increases, the liquid expands. Standard liquids are mercury or alcohol. Another simple method for temperature measurement is the bimetal thermometer made from two bonded sheet metals. When temperature increases, the different coefficient of thermal expansion will generate a thermo-mechanical stress. The resulting bending is a measure for the temperature. Less common, but very important for basic metrology is the gas thermometer: When an ideal gas is heated and volume is kept constant, the pressure increases. In this way, measuring temperature is reduced to measure pressure. For practical application, this is very difficult, in fact, but gas thermometers are very important as standards and for calibrating other sensors.

For sensors, we need an effect that converts the temperature into an electric signal. In the following sections, we will discuss the following three effects:

- Thermoelectric effect
- Thermoresistive effect
- Pyroelectric effect

1.1.1 Thermoelectric Effect (Seebeck Effect)

In 1821, Thomas Johann Seebeck found that electric power can be generated by an arrangement of two metals. The classical configuration is shown in Figure 1.1. We join two lines of metals A and B; as an example, this might be copper versus iron. In a second junction, we go back to copper. We measure voltage between the two copper lines U, which is called thermoelectric

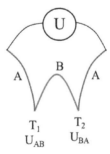

Figure 1.1 Thermoelectric effect between two metals A and B.

voltage or Seebeck voltage. For simplicity, we assume that our voltmeter has an infinitely high entrance resistance.

We assume that the voltage must be generated at the joining points, and we split it into two contact voltages, U_{AB} and U_{BA}. For the moment, we will neglect nonlinear effects and write the voltage as a linear function of temperature, which depends on the two metals A and B:

$$U = U_{AB} + U_{BA} = k_{AB}T_1 + k_{BA}T_2 \tag{1.2}$$

For zero temperature difference, $T_1 = T_2$, the voltage must be zero; otherwise, we would create energy, which cannot happen. Therefore:

$$k_{AB} = -k_{BA} \quad \text{and} \tag{1.3}$$

$$U = k_{AB}(T_1 - T_2). \tag{1.4}$$

where k_{AB} is the linear thermoelectric coefficient, also called linear Seebeck coefficient.

We make another thought experiment and introduce a third metal R (for Reference) in between metals A and B, as shown in Figure 1.2:

$$U = U_{AB} + U_{BR} + U_{RA} = k_{AB}T_1 + k_{BR}T_2 + k_{RA}T_3 \tag{1.5}$$

Again, for $T_1 = T_2 = T_3$, the voltage must vanish:

$$k_{AB} + k_{BR} + k_{RA} = 0 \tag{1.6}$$

$$k_{AB} + k_{BR} - k_{AR} = 0, \quad \text{or} \tag{1.7}$$

$$k_{AB} = k_{AR} - k_{BR}. \tag{1.8}$$

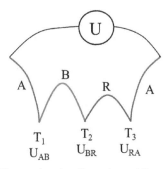

Figure 1.2 Thermoelectric effect when adding a third metal R.

Table 1.1   Thermoelectric voltage series

Metal	Thermoelectric Coefficient [μV/K]
Bismuth	−72
Constantan	−35
Nickel	−15
Platinum	0
Aluminum	3,5
Rhodium	6
Copper	6,5
Gold	6,5
Antimony	47

This is an important finding because it allows expressing the thermoelectric coefficient between the two metals A and B by the two coefficients of A versus R and B versus R. Imagine we want to characterize the effect for 10 metals. Then, we would set up a matrix of 10 times 10 metals and we would have to measure 90 coefficients. Equation (1.8) makes it simple: we define one metal as the reference metal. Traditionally, this is platinum. Then, we have to conduct nine experiments, measuring every metal versus platinum, and the rest can be calculated. The list of thermoelectric voltages versus Pt is called the thermoelectric voltage series. Table 1.1 is an extract of this series.

Let us focus once more on energy: zero temperature difference makes zero voltage. This is a consequence of the law of the conservation of energy. When we maintain a temperature difference between junctions 1 and 2, there will be heat flowing between the junctions, and we bring energy in the system. Part of this energy, a very small part, in fact, is regained as electric energy. Therefore, we see that thermoelectricity is an energy transformation: from thermal energy to electric energy. In fact, it works the other way round, too: if you drive a current through the deployment of Figure 1.1, then one of the junctions will heat up and the other will cool down. This is called the Peltier effect, which can be used for electric cooling of small items.

To date, our discussion has been purely phenomenological; we have been describing the effect. But why is there a thermoelectric voltage at all? To understand the reason behind, we have to look at the energies of the electrons that transport electric current in metals. First, we consider two metals separately, without contact and without thermal movement of the particles. The electrons fill all the possible energy levels, from smallest energy onward. The highest energy level filled is called the Fermi level. But the metals have a temperature, and hence there is thermal movement. Therefore, some charge

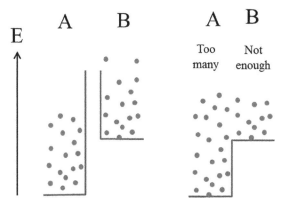

Figure 1.3   Explanation of the thermoelectric effect between two metals A and B. Left: metals not in contact; right: metals in contact, and charge carriers can be exchanged between A and B.

carriers move above the Fermi level, making an electron cloud that gets thinner at higher energy, as shown in Figure 1.3. The Fermi level depends on the material, the type of atoms involved, and the way they are connected in a crystal lattice. Therefore, the two metals A and B have different Fermi levels, and the electron clouds are at different heights on the energy scale, as shown in Figure 1.3.

Now, we make contact and electrons can move from metal A to B and vice versa. Metal B has a higher Fermi level, so electrons will just move downhill from B to A. This movement is driven by the Fermi energies. In this way, the difference in potential increases with accumulation of charges in metal A. The difference in potential is an electric field, and an electron in a field is subject to a force, the Coulomb force, which acts in a way to reduce the field. Thus, it will drive the electrons back right. Coulomb force increases with increasing potential, and there will be an equilibrium point when the two forces cancel. This equilibrium determines the contact force. What happens when the temperature rises? The thermal movement of the electrons increases, and in this way, the electron cloud above the Fermi level will change. Also, the Fermi energy may change with temperature. Finally, there will again be an equilibrium, but at another potential, and we understand that the contact voltage is a function of temperature. Finally, we look at the thermoelectric voltage, which is the sum of the contact voltages. Being opposite in sign, for $T_1 = T_2$, the two contact voltages just cancel. For $T_1 \neq T_2$, they do not cancel any more and a net thermoelectric voltage is observed.

We see a number of metals in Table 1.1, but where is silicon? In fact, it can be very high or very low, depending on the doping. Also this can be understood regarding Fermi levels[2]. In semiconductors, charge carriers may be electrons or holes, which are defective electrons. Most of them are fixed in a valence band. Only those in the conduction band are movable and contribute to electric transport. Now, the number of electrons in the conduction band increases significantly with increasing temperature. As an effect, small changes in temperature will cause large changes in electric behavior, such as thermoelectricity. Some typical values for a medium doping ($10^{16}/cm^3$) are -1200 µV/K for n-doped and $+1300$ µV/K for p-doped monocrystalline silicon. The thermoelectric constants in polycrystalline silicon are lower than those in single-crystal material, but still high with respect to metals. Typically, we expect values around 200 µV/K for the thin-film material we use for thermal sensors. The highest values are measured for low doped material. With increasing doping, thermoelectric voltage reduces. Finally, for very high doping, semiconductors will become conductive, which is called metallic degeneration. The thermoelectric voltage will then decrease to the small values typical for metals.

Practically, thermocouples are available as readymade sensors with metal shielding. Often used combinations are:

K-type: NiCR/NiAl,
J-type: Fe/CuNi, and
S-Type: Pt/PtRh (10%) for high temperature up to 1600 °C.

Thermocouples generate a voltage; therefore, line resistivity does not disturb measurement as long as the voltage is measured with high impedance. Thermocouples always measure a temperature difference! When you want to measure absolute temperature, you need a reference temperature at one of the junctions, which is called a "cold junction". In the laboratory, ice water is a good reference for 0 °C, but, in fact, not in technical implementation. For standard material combinations, such as the K-type, you can get a readymade "electronic cold junction". This is a temperature-controlled voltage source, which exactly simulates the contact resistance of the K-type NiCr/NiAl contact. But then, this electronic circuit needs also a measurement of absolute temperature. It may use a resistive thermometer as described in Chapter 2.

Shielded thermocouples are reliable and robust, but slow, since it takes some seconds for a temperature signal to travel through the metal housing.

---

[2]http://www.iue.tuwien.ac.at/phd/knaipp/node18.html

When you need a small and fast device, you can make a thermocouple yourself from two thin wires. Bifilar wires with two different metals are commercially available.

Another method to make thermocouples is thin-film technology. In fact, for sensors, this is the most interesting option. When we make a thermocouple from two thin films with a thickness of approximately 300 nm, we can get thermal response times in the range of milliseconds. Very often, sensor technologists apply many thermocouples in a series to increase voltage. This is called a thermopile. What material would we use for thin films? Looking at the metals in Table 1.1, we would like to combine a material from the top of the list with the one from the bottom. Consequently, a classic solution is the bismuth–antimony thermopile, which achieves a high voltage compared to all other metal combinations. A problem is technology: bismuth and antimony are both poisonous, and so you will end up in buying dedicated machines for this process in order not to contaminate your equipment. For this reason, polysilicon is widely used. It has a high thermoelectric coefficient of approximately 200 μV/K, and its deposition is a standard process. Disadvantages are the high electric resistivity of the thin films and the expensive equipment for the low-pressure chemical vapor deposition process used to make polysilicon, which we will discuss in Chapter 2. When polysilicon is one material, what will be the second? A classic combination is polysilicon versus aluminum[3]. This can be fabricated by a standard processes, but the aluminum restricts us to low-temperature isolation layers for final passivation. For this reason, polysilicon versus tungsten–titanium has been developed[4]. TiW withstands high temperatures and allows using high-temperature deposited films such as LPCVD (low-pressure chemical vapor deposition) for passivation.

1.1.2 Thermoresistive Effect

The resistivity of a material is a function of temperature:

$$R = R_0(1 + \alpha\Delta T + \beta\Delta T^2 + \dots) \tag{1.9}$$

---

[3]Infrared thermopile sensors with high sensitivity and very low temperature coefficient. J. Schieferdecker, R. Quad, E. Holzenkämpfer, M. Schulze. Sensors and Actuators A, Vol. 47 (1995).

[4]Buchner, R., C. Sosna, M. Maiwald, W. Benecke and W. Lang: A high-temperature thermopile fabrication process for thermal flow sensors. Sensors and Actuators A: Physical 130–131, 262–266 (2006).

where α is the linear thermoresistive coefficient and β is the quadratic coefficient. Differently from thermocouples, thermoresistive thermometers measure absolute temperature.

In metals, the resistivity increases with temperature ($\alpha > 0$). This is called a PTC (positive temperature coefficient) material. Typical values for metals are in the range of 0.4–0.8%/K. The most important materials are:

Platinum: $\alpha = 3.8 \times 10^{-3}/K$; $\beta = 0.6 \times 10^{-6}/K^2$ and
Nickel: $\alpha = 6{,}7 \times 10^{-3}/K$; $\beta = 9 \times 10^{-6}/K^2$.

We see that nickel is more sensitive, but platinum is more linear. Examples for metal thermometers are Pt100, a platinum resistor with a base resistivity of 100 Ω at 0 °C, and Ni1000, a nickel resistor with a resistivity of 1000 Ω at 0 °C. These thermometers can be produced as metal wires, in thick-film technology or in thin-film technology. The advantage is the high precision. Disadvantages are a limited temperature range, Ni up to 200 °C, Pt up to 800 °C, and large size. Furthermore, self-heating: to measure the resistivity, we apply a measurement current, and this generates Joule heating. We have to be careful to apply small measurement currents not to distort what we want to measure.

Low-doped semiconductors have a negative TCR (temperature coefficient of electrical resistance), since at elevated temperature, more electrons move from the valence band to the conduction band. Medium- and high-doped semiconductors change to positive TCR; at very high doping, they behave like a metal and show a small positive TCR (metallic degeneration).

There are some exotic materials specially made for extremely high-temperature coefficient, which are called thermistors. Negative temperature coefficient materials (NTCs) are made from metal oxides. Temperature lifts electrons into the conductive band. The temperature coefficient is 10 times as high as for metals, but they show very high nonlinearity. They are used to make small resistive temperature sensors.

Some ceramic materials have a positive TCR up to 100%/K. Within a few degrees, these PTC materials change from low to high resistivity. Applications for measurement are rare, but PTCs are used as temperature switches and for securing electronic circuits against overcurrent.

For ASIC technology, we would also like to have thermometers in silicon. When we want to protect an ASIC against overheating by a security circuit, we must measure temperature on the chip. Another example is a preamplifier for a sensor. When the circuit knows the temperature cross-sensitivity of the sensor and the system temperature, it can perform numerical temperature compensation. To measure the temperature on silicon, we can use the temperature dependence of the characteristics of a diode:

$$I = I_S e^{\frac{e_0 U}{k_B T}} \qquad I_S = \text{Saturation current} \qquad (1.10)$$

We write voltage as a function of T:

$$U = \frac{k_B}{e_0} \left(\ln \frac{I}{I_S} \right) T \qquad (1.11)$$

The problem is that the saturation current I_S depends on doping and will spread between several copies of the same type of diode. Therefore, we want to remove I_S from Equation (1.10), which can be done by measuring with two different currents I_1 and I_2:

$$U_1 - U_2 = \frac{k_B}{e_0} \left(\ln\frac{I_1}{I_S} - \ln\frac{I_2}{I_S} \right) T = \frac{k_B}{e_0} \left(\ln\frac{I_1}{I_2} \right) T \qquad (1.12)$$

In this way, finally, I_S is removed from the equation. To implement this, we use only one diode and switch the current between I_1 and I_2.

1.1.3 Pyroelectric Effect

Some dielectric materials have a natural dipole moment (like a permanent magnet has a natural magnetization). This dipole moment will change if the material is deformed (piezoelectric effect) or if the temperature is changed (pyroelectric effect). In crystals with high symmetry, all directions are equivalent, which forbids a permanent dipole moment, and therefore, pyroelectric and piezoelectric effects are forbidden in high-symmetry material.

Why does the effect only show by materials that have a natural dipole moment? The symmetry argument runs as follows:

- The effect generates a voltage
- The voltage has a sign: positive or negative
- When I heat, where should the positive sign point to? I need a specific direction anchored in the microstructure of the material
- This direction can be given by a permanent dipole moment in the material.

Table 1.2   Effects to measure the temperature

Effect	Measurand		Response	
Thermoelectric	ΔT	Temperature difference	U	Voltage
Thermoresistive	T	Absolute temperature	R	Resistivity
Pyroelectric	dT/dt	Change of temperature with time	Q	Charge

Is silicon pyroelectric? It has a cubic symmetry. When I invert the cube, the lattice does not change. This means that the symmetry operation of inversion (x→−x; y→−y; z→−z) is part of the crystal symmetry. When I heat a piece of silicon, no matter where I put the electrodes to pick up a possible voltage, applying inversion, I can show that U =−U. Therefore, U must vanish. For the same reason, silicon cannot be piezoelectric. What materials remain? Single-crystalline quartz is SiO_2 like glass, but has a trigonal symmetry; pyroelectric and piezoelectric effects do show. Some ceramics like $BaTiO_3$ (barium titanate) and some polarized thin films show pyroelectric and piezoelectric effects.

What does the pyroelectric effect actually measure? When I heat the crystal, charges accumulate at the surface, which I can pick up and measure with a voltmeter. After some time, due to ions in the air and creeping currents on the surface, the charges will disappear. When I change temperature again, I will find a signal again. The effect actually does not go with temperature, but with the change of temperature in time. Table 1.2 sums up the effects to measure temperature.

1.2 Radiation Thermometer

The family of thermal sensors is numerous, and we have to pick an example to discuss in detail. I chose the thermopile sensor for infrared radiation since it is interesting, easy to model and practically important.

Infrared sensors are used for non-contact temperature measurement. When a body is hot, it emits infrared radiation (IR, "heat radiation"). A pyrometer (pyr is the Greek word for fire) measures the IR to calculate the temperature.

First, we have to understand the emission. Each body emits thermal infrared radiation with a power P_{IR} according to the area A, emissivity ε_R, and absolute temperature T:

$$P_{IR} = const \cdot A \cdot \varepsilon_R \cdot T^4 \tag{1.13}$$

The emissivity ε_R is a property of the surface. Luckily, for infrared, most surfaces have an emissivity close to 1, so we do not have to discuss ε_R in detail. The one big exception is metals that reflect. When we look at a metal mirror with the pyrometer, we do not see the temperature of the mirror, but we look at whatever the mirror shows us.

When temperature increases, two things happen: first, the infrared radiation power increases with a power of 4 of the temperature. Then, the emitted radiation shifts to a shorter wavelength. Figure 1.4 shows the electromagnetic spectrum of surfaces at different temperatures. At room temperature, the center of the wavelength is approximately 10 μm in the medium-infrared region. We cannot see this, but feel it indirectly with the temperature sensors in our skin. The wolfram filament of an electric lamp is at around 3000 K. There, the center of the wavelength is at 1 μm and the emitted power per area increases by 4 orders of magnitude. The visible part of the spectrum is from 380 nm (violet) to 780 nm (red). This way, at room temperature, the part of emission in the visible region is negligible, and at 3000 K, it is considerable,

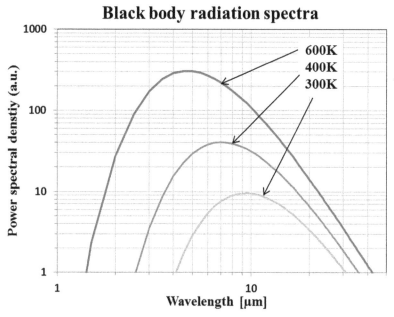

Figure 1.4 Radiation spectrum of a blackbody. For higher temperature, the power density increases and the peak shifts to shorter wavelength. For pyrometry at room temperature, we need a sensor for infrared radiation with a wavelength of about 10 μm.

but yet not 50%. Hence, we move away from filament lamps to more efficient light sources like LEDs.

For non-contact thermometry, we are interested in room temperature up to approximately 1000 K, so our sensor must be sensitive from 10 μm down to 1 μm. An example is the ear fever thermometer. It measures the IR radiation emitted from the skin in the ear and calculates the body temperature.

How can we use a temperature sensor to measure IR radiation? The quantity to be measured is radiation, and the effect this makes is a small amount of heat. How do we achieve a temperature increase as large as possible from a small amount of energy? The answer is our choice of a transduction structure, which is a thin membrane with a measurement spot and thermal sensors integrated on the membrane. Using a thin film of silicon nitride for the membrane, the thermal capacity of the measurement spot will be low, and the thermal insulation from the bulk of the chip is appreciable. In this way, little energy will cause large temperature increase.

This type of sensor is called a thermopile. A cross section of the sensor is shown in Figure 1.5, and a top view in Figure 1.6. In the thermopile, we use a membrane from very thin (300 nm) silicon nitride. The IR radiation is absorbed in the measurement spot. This generates a small temperature difference (ΔT) between the measurement spot and the bulk of the chip. To measure ΔT, we use a number of thermocouples with one end under the measurement spot, and the other end on the bulk. These are made from polysilicon versus aluminum, to use the high thermoelectric coefficient of the semiconducting material. We choose the thermoelectric effect for measurement because we are interested in temperature difference, neither in absolute temperature nor in temperature changes with time. The IR thermopile will be the paradigm example of this book to discuss sensor principles and sensor technology. Figures 1.7 and 1.8 give examples of commercial thermocouple devices.

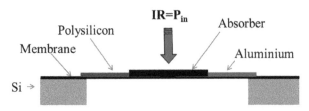

Figure 1.5   Thermopile cross section.

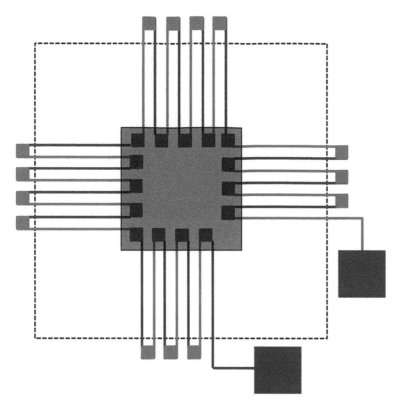

Figure 1.6 Thermopile top view.

Figure 1.7 Commercial thermopiles: single sensors in TO metal housing and thermopile arrays. Picture by permission of Heimann Sensor GmbH[5].

---

[5]https://www.heimannsensor.com/products_imaging.php.

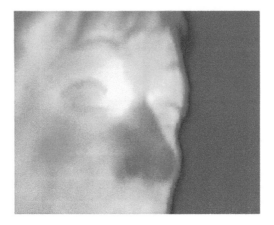

Figure 1.8 A human face seen by a thermopile array of 80×64 pixels. The pitch between two pixels is 90 μm. Presented with permission from Heimann Sensor GmbH.

1.2.1 Static Modeling of Thermal Transport and Understanding Sensitivity

To understand the function of a thermopile, we have to model thermal transport of the membrane. First, we look at an electric equivalent shown in Figure 1.9.

A simplified model for the IR thermopile looks at the power flowing in and out of the membrane.

	Thermal	Electric Equivalent
	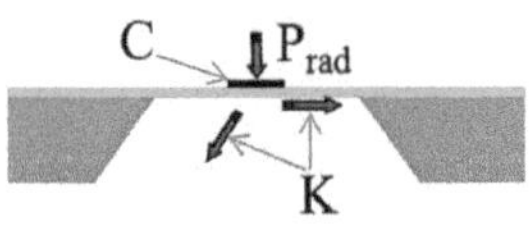 C $\downarrow P_{rad}$ K	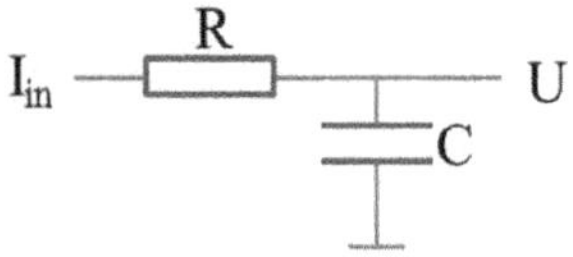 I_{in} —[R]— U C
Input	Infrared radiation power P_{rad}	Current in I_{in}
Capacity	Thermal capacity C	Electric capacity C
Determining the flow out	Thermal conductivity of the membrane K	Electric conductivity 1/R
Measured value	Temperature difference measurement spot and bulk	Voltage U
Time constant	Thermal time constant C/K	RC

Figure 1.9 A thermopile sensor and an electric equivalent circuit.

The power flowing in the membrane is the infrared radiation power P_{rad}, which is absorbed in the measurement spot and heats the membrane.

The power flows out in three ways:

1. Heat conduction through the interconnects and through the membrane.
2. Wall heat transfer into the surrounding gas and convective transport.
3. Radiative power emission by the membrane.

The latter, radiative emission by the membrane, can be neglected. The difference from room temperature is very small, $\Delta T << 1\,°C$. Radiative emission goes with T power of 4, so it can be neglected for a small ΔT.

For <u>heat conduction</u>, we first analyze the lines[6]. Thermal conductivity through a single line is described by:

$$P_{line} = -\frac{A_l}{l} k \Delta T \tag{1.14}$$

where A_l is the cross section of the line given by the product of the film thickness d (typically 300 nm) and the line width w (typically 10 μm). Concerning the sign, we follow the convention that positive energy flow is into the measurement spot. The membrane is heated slightly; heat will be flowing out and hence P will be negative.

There are n lines from polysilicon and n lines from metal. Heat conduction through all the interconnects is:

$$P_{cond} = -n\frac{A_l}{l}(k_{Si} + k_M)\Delta T \tag{1.15}$$

What about the heat conduction in the membrane? The thermal conductivity of silicon nitride is 2.5 W/mK, very low compared to the conductivity of polysilicon (150 W/mK) and aluminum (230 W/mK). From this, we conclude that the energy flow through the membrane is much lower than the energy flow through the interconnects and we decide to neglect the membrane in the energy balance. Using the specific figures of our example sensor, as given in Table 1.3, we find for the conductive heat transport of the thermopile:

$$P_{cond} = -170\,\frac{\mu W}{K}\Delta T \tag{1.16}$$

<u>Wall heat transfer</u> is the convective transport of energy from the membrane to the surrounding gas. In fact, convective transport is a difficult problem

---

[6]D. Pitts and L. Sissom: Heat Transfer; Schaums Outline Series, McGraw Hill.

Table 1.3   Parameters of the thermal model of a thermopile

Symbol	Parameter	Specific Value of Our Paradigm Example
C	Thermal capacity of the measurement spot	2.0 µJ/K
α	Wall heat transfer coefficient	100 W/m²K
	Size of the measurement spot	260 µm × 260 µm
	Size of the membrane	1 mm × 1 mm
A	Heated area of the membrane	0.6 mm × 0.6 mm
A_1	Cross section of the line	10 µm × 300 nm
L	Length of interconnects	350 µm
n	Number of thermopiles	52
k_{Si}	Thermal conductivity of silicon	150 W/mK
k_M	Thermal conductivity of metal	230 W/mK
K	Thermal conductivity of the device	
k_{SiAl}	Thermoelectric coefficient polysilicon vs aluminum	200 µV/K
ΔT	Temperature difference between membrane and bulk silicon	

of thermodynamics due to nonlinear effects induced by free convective flow around a heated body. However, knowing that $\Delta T \ll 1\,°C$, we assume that there will be no convective flow induced. This case is called "stagnant flow", and wall heat transport becomes linear with ΔT:

$$P_{conv} = -2\alpha A\Delta T \qquad (1.17)$$

This linear approach for wall heat transfer is known as "Newton's law of cooling". It is obvious that the power flow is proportional to the heated area of the membrane A. What area should I use? The central measurement spot is certainly heated, and the rim just close to the bulk is certainly not heated. I choose a line in the middle between the measurement spot and the bulk, and hence I assume $A = 0.6\ mm^2$. The factor 2 reflects the fact that the membrane has two sides. Take care not to mix A (heated area of the membrane) with A_I (cross section of the interconnects)! The coefficient α is called the wall heat transfer coefficient. For small membranes, it has been measured to be 100 W/m²K. Experiments to verify the thermal transport of sensor membranes can be carried out by heating up a membrane and plotting temperature versus input power. To separate conduction and convection, the experiment is conducted in air and vacuum, since in vacuum, convection vanishes while conduction does not[7].

Again we use the figures from the example and find:

$$P_{conv} = -70 \frac{\mu W}{K} \Delta T \qquad (1.18)$$

We learn that heat loss is primarily due to the conduction in interconnects.

The total power flowing out of the membrane is described by K, the thermal conductivity of the device:

$$P_{out} = P_{conv} + P_{cond} = -[2\alpha A + n\frac{A_I}{l}(k_{Si}+k_M)]\Delta T = -K\Delta T \quad (1.19)$$

$$K = 2\alpha A + n\frac{A_I}{l}(k_{Si} + k_M) \qquad (1.20)$$

In our example, K = 70 µW/K + 170 µW/K = 240 µW/K.

Please note that K designates the thermal conductivity of the device (the membrane), while k stands for the thermal conductivity of a material. This is like R for the resistivity of a device and ρ for the specific resistivity of a material.

We assume that the IR radiation is absorbed to 100%, thus P_{in} is the radiation power.

The difference of power in and power out will change the temperature of the membrane:

$$P_{in} + P_{out} = C\Delta \dot{T} \qquad (1.21)$$

First, we calculate the sensitivity in steady state:

$$\dot{T} = 0; \qquad P_{in} + P_{out} = 0 \qquad (1.22)$$

$$\Delta T = \frac{P_{in}}{K} \qquad (1.23)$$

This gives us the temperature difference generated by the radiation. To understand the sensitivity of the sensor, we must correlate this with the thermoelectric voltage to obtain the voltage induced by radiation. The thermoelectric voltage U is given by the thermoelectric coefficient:

$$U = nk_{SiAl}\Delta T \qquad (1.24)$$

---

[7]Heat Transport from a Chip. W. Lang, IEEE Trans. on Electron Devices, Vol. 37, No. 4 (1990) S. 958–963.

The sensitivity E is the voltage U generated by a radiation power P_{in}:

$$E = \frac{U}{P_{in}} = \frac{nk_{SiAl}}{K} \tag{1.25}$$

For our example, we calculate E = 43 V/W. Data sheets of comparable devices[8] range between 44 and 58 V/W. From this, we conclude that our simple model describes the sensor quite accurately.

1.2.2 Modeling Dynamic Behavior

Now, we have understood the steady-state response. But how does the sensor react on a change in radiation? To analyze the time behavior, we have to look at the full Equation (1.21). First, we investigate step response. Initially, there is no IR radiation and the sensor is in thermal equilibrium with the surrounding at $\Delta T = 0$. Then, at $t = 0$, the radiation P_{in} is switched on with a step function. The membrane is heated up, but it needs some time to reach the new equilibrium. The electric analogue (Figure 1.9) is a lowpass realized with an RC circuit fed with a square wave. Equation (1.21) is a first-order differential equation, and we can solve it using an exponential ansatz:

$$\Delta T(t) = \Delta T_0 \left(1 - e^{-\frac{t}{\tau}} \right) \tag{1.26}$$

We find:

$$P_{in} = C\Delta\dot{T} + K\Delta T = \frac{C}{\tau}\Delta T_0 e^{-\frac{t}{\tau}} + K\Delta T_0 - K\Delta T_0 e^{-\frac{t}{\tau}} \tag{1.27}$$

This equation has time-dependent and time-independent terms. The solution must hold for all points in time; therefore, the time-dependent parts must be equal and the time-independent parts must also be equal. This gives us two equations.

$$\begin{aligned} (1) \quad & P_{in} = K\Delta T_0 \\ (2) \quad & \tau = \frac{C}{K} \end{aligned} \tag{1.28}$$

(1) is the solution for the sensitivity in the steady state we already know. Sensitivity depends only on the conductivity K.
(2) describes the thermal time constant τ as the ratio of the thermal capacity C to the thermal conductivity K.

This behavior is called first-order time response. It is described by a delay time constant τ. At time τ after the step, the sensor output has reached

---

[8]http://www.heimannsensor.com/Datasheets/Datasheet-2b-TO18_rev2.pdf.

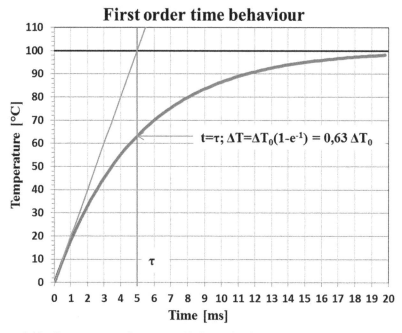

Figure 1.10 Step response of a sensor with first-order time response. The step is from 0 °C to 100 °C.

63% of its final value, as shown in Figure 1.10. This can be calculated from Equation (1.26) by:

$$\Delta T(\tau) = \Delta T_0 \left(1 - e^{-1}\right) = 0,63 \,\Delta T_0 \tag{1.29}$$

For this reason, the time constant we use is also called the "63% time constant". This relation is useful to determine the time constant from a measured step response. Another way to determine τ is to draw a tangent in the origin. It will cross the ΔT_0 value at time τ.

For our example sensor, we calculate a thermal capacity of 2.0 μJ/K and a time constant of 8.3 ms. Typical measured values for thin-film thermopiles range from 6 to 20 ms.

It is also interesting to analyze the sinus response of a first-order system. The graph of the Bode plot in Figure 1.11 gives the amplitude and the phase of the response curve. The calculus is done in every textbook of basic electric engineering and must not be repeated here. The essential messages are: There is a cutoff frequency f_0, which is the inverse of the time constant τ. Below the cutoff, the amplitude is constant. Above the cutoff, the amplitude falls with

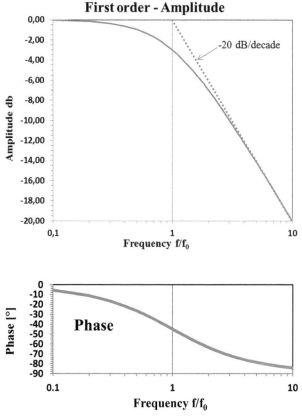

Figure 1.11 Amplitude and phase of a first-order system.

a slope of -20 db/decade. This means that a 10 times increase in frequency makes 1/10 of amplitude. The phase lag increases with frequency. At cutoff frequency, the phase is $-45°$. At high frequency, it approaches $-90°$.

If the measured phenomenon changes slowly with respect to the time constant τ, then the sensor will give realistic information about the phenomenon. If not, the slowness of the sensor will cause the following three deviations between the physical phenomenon and the sensor output:

1. Phase shift: The electric signal will be a little later than the measured physical effect.
2. Reduced sensitivity: The electric signal will not reach the full amplitude.
3. Signal distortion: A rectangular signal will be distorted in the form of a "shark fin".

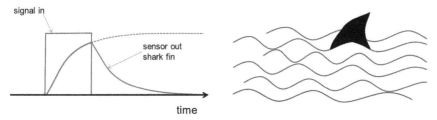

Figure 1.12 When the time constant of the sensor is not small with respect to the signal, a rectangular signal will be distorted in the form of a "shark fin".
Three deviations occur: (1) the signal is smaller; (2) it comes later; and (3) it is distorted.

Table 1.4 Trade-off between sensitivity and bandwidth

	For High Sensitivity	For High Bandwidth
Thermal conductivity K	Should be small	Should be large
Thermal capacity C	No impact	Should be small

Most measurement systems show first-order time response. Second-order time response occurs if a system stores energy, and therefore may vibrate, such as an LC circuit or a mechanical spring mass system. We will discuss this behavior in the chapter on accelerometers, which are spring mass systems.

What do we learn for the layout of a thermopile?

The sensor should be sensitive and fast at the same time. Thermal capacity must be as small as possible. This makes the sensor fast and does not influence the sensitivity. Therefore, we use a very thin membrane, resulting in small mass and small heat capacity. Concerning the thermal conductivity, there is a trade-off between sensitivity and time constant. If the membrane is very well isolated (large, thin, and vacuum-encapsulated), then K will become small. This renders the device sensitive, but also makes it slow.

From Table 1.4, we conclude the following: Thermal capacity should be as small as possible for the bandwidth. Then, a good thermal isolation should be achieved to obtain sensitivity. Improving bandwidth by increasing the thermal conductivity should be avoided, since it ruins sensitivity. Knowing this, we may reconsider our choice of silicon nitride as a membrane material. Nitride films can be made as thin as 300 nm. Typically, silicon membranes are approximately 10 μm thick, so their thermal capacity will be 30 times higher than that of nitride membranes. Furthermore, thermal conductivity of the insulating silicon nitride is low (k ≈ 2.5 W/mK), whereas that of the semiconducting silicon is high (k ≈ 100 W/mK). Therefore, silicon nitride is definitively the material of choice.

1.2.3 The Noise of the Thermopile

The given model allows us to estimate the sensitivity of the thermopile. However, sensitivity can be increased just by an amplifier and hence it does not say much about the quality of a sensor. The most important figure for the quality of a sensor is resolution. Resolution is often limited by noise. Therefore, we have to estimate the noise of the thermopile. Here, we follow the program defined in the introduction: understand the physics first, calculate the sensitivity, and then calculate the noise.

The noise of the thermopile is thermal noise from the resistivity of the sensor. A particle is never totally at rest. Only at the absolute zero temperature (0 K) it would not move, but this never happens. All particles, such as atoms and electrons always show a small thermal movement. The higher the temperature, the larger is the thermal movement. Temperature is just another word for the average thermal kinetic energy of the particles. Also the electrons in the resistor will show thermal movement. Moving charges make an electric current, so there will be a noise current. Nyquist[9] proposed to consider noise current as a superposition of vibrations in the resistor. Each of those vibrations will have an average thermal energy:

$$E_N = k_B T \tag{1.30}$$

where k_B is the Boltzmann constant. Half of the average energy is kinetic, half of it is potential energy. There will be a fundamental frequency at f_0, the first harmonic at $2f_0$, the second at $3f_0$, and so on. In this way, the energy is equally distributed over the frequency spectrum, and we can write the spectral noise density as follows:

$$E_N \delta f = k_B T \delta f = \delta P_N. \tag{1.31}$$

We write the noise density for power, since energy multiplied with frequency equals energy over time, which is power. When we look at a frequency band with a bandwidth Δf, we can integrate this and write:

$$P_N = k_B T \Delta f. \tag{1.32}$$

Now, we focus on the electric side. Our question to find noise voltage U_N is: What voltage would it need to generate the noise power described by the thermal movement in Equation (1.32)? To correlate this voltage to the

---

[9]H. Nyquist: Thermal agitation of electric charge in conductor. Physical Review, Vol. 12 (1928)

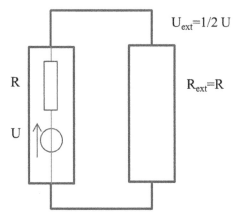

Figure 1.13 Nyquist's consideration for noise power. Voltage generation source and resistivity are spread over the resistor. We assume impedance matching $R_{ext} = R$.
External voltage is $U_{ext} = \frac{1}{2}U$, Current is $I = \frac{U}{2R}$, External power is $P = UI = \frac{U^2}{4R}$.

noise power, we regard the resistor as a voltage source U with an internal resistance R. Power is extracted by an external resistor R_{ext} as shown in Figure 1.13. For power extraction, we assume impedance matching $R_{ext} = R$. Then, we get externally a power of

$$P_N = \frac{U_N^2}{4R} \qquad (1.33)$$

This gives:

$$\frac{U_N^2}{4R} = P_N = k_B T \Delta f \qquad (1.34)$$

$$U_N = \sqrt{4k_B R T \Delta f} \qquad (1.35)$$

This is called Nyquist's (or Johnson's) formula for the thermal noise of a resistor. It is important to note that this voltage is also a time average value like noise energy and noise power.

I would like to discuss the essential point again from a more practical viewpoint. Energy and power are correlated by time. What time should we apply to correlate between the thermodynamic argument for noise energy in Equation (1.30) and the electric argument for noise power in Equation (1.33)? A reasonable time would be the time dt we allow to accumulate data for each single measurement. This time will certainly determine the noise. Long measurement time means many data, better statistics, and therefore less noise. The measurement time is the inverse value of the measurement bandwidth Δf.

When we allow long measurement time, we get many data per point, and in this way, better statistics and less noise, but it results in small bandwidth and slow measurement. For example, you can reduce electronic noise using an RC lowpass filter, but only at the cost of reducing bandwidth. This is well known from the treble filter in radios: they not only reduce noise, but also bandwidth and thus the sound will become a little "muddy".

A noise which is equally distributed over frequency is called white noise. When we increase bandwidth, there is more noise power. When we assume infinite bandwidth, is there infinite noise power? Of course, not. Infinite bandwidth is an unrealistic assumption. Every sensor and every electronics has an upper frequency limit. What we should bear in mind is that white noise is a simplified assumption which will not hold if we look at very high frequencies.

The formula also directly tells us what we can do to reduce noise. The meaning of the terms is:

T: The noise increases with temperature, since the movement of the electrons increases with temperature. Thus, the noise can be reduced by cooling the sensor. This is done in laboratory equipment. When thermopiles are used for IR spectroscopy or astronomy, they are cooled with liquid nitrogen to 77 K.

R: Due to Ohms law, the noise increases with the resistivity of the sensor. Unfortunately, the thermopile has a high internal resistivity due to the interconnects in thin-film technology and due to the many contacts between the two metals. A typical value is 50 kΩ. Therefore, we would like to make sensors with less resistance to reduce Nyquist noise. We could reduce the number of thermopairs in a thermopile, let us say, in half (1/2). The noise would reduce to $1/\sqrt{2}$ = 71%, but the sensitivity would reduce to 50%. Obviously, it does not pay out.

Δf: The noise increases with the bandwidth of measurement. To reduce noise, the bandwidth can be reduced using a filter (like the "treble" filter in an audio amplifier). There is a trade-off between noise and response time. Small bandwidth reduces noise, but increases response time. An example is the accelerometer of an airbag system. To reduce the noise level, we can apply a lowpass filter, e.g., 0–100 Hz. But then, regarding Figures 1.11 and 1.12, we have to expect ten microsecond time delay between the crash and the signal. This makes the airbag useless. The sensor signal must be available faster than 250 µs after the crash. This means that any filter setting below 4000 Hz will render the sensor too slow for this application. For this reason, automotive companies normally ask for 4 kHz bandwidth for airbag accelerometers.

Let us go back to the thermopile example: The resistivity of the device is 50 kΩ. We assume a bandwidth of 10 Hz, which is high enough for temperature measurement. Then, the noise is calculated as:

$$U_{\text{Noise}}(0\ldots10Hz) = \sqrt{4k_B T R \Delta f} = 90 \text{ nV} \qquad (1.36)$$

This is important information, but not yet what we need. To get information on the resolution, we have to correlate U_{Noise} with the sensitivity E. In this way, we define the noise equivalent power NEP (Δf):

$$NEP(\Delta f) = \frac{U_{\text{Noise}}(\Delta f)}{E} \qquad (1.37)$$

In the case of the thermopile, the sensitivity is $E = 43$ V/W and

$$NEP(0\ldots10Hz) = \frac{90nV}{43V/W} = 2,1 \text{ nW} \ (0\ldots10 \text{ Hz}). \qquad (1.38)$$

To compare different sensors, it is more useful to choose a form that is independent of bandwidth. We achieve this by dividing by the square root of bandwidth:

$$NEP = \frac{U_{\text{Noise}}(\Delta f)}{E\sqrt{\Delta f}} = \frac{90nV}{43V/W\sqrt{10Hz}} = 0,7 \frac{nW}{\sqrt{Hz}} \qquad (1.39)$$

Noise equivalent signal is one of the most important figures to characterize a sensor. Here, we calculate noise equivalent infrared power. Later, we will also discuss noise equivalent acceleration of accelerometer and others. Whenever you see this strange kind of dimension $1/\sqrt{Hz}$, you know that noise is discussed and it is given in an bandwidth-independent form.

What values do real sensors achieve? The following are the examples

T11262 by Hamamatsu:	$E = 50$ V/W; NEP = 0.9 nW/$\sqrt{Hz}$,	$\tau = 20$ ms.[10]

or the

HMS by Heimann sensor:	$E = 58$ V/W; NEP = 0.65 nW/$\sqrt{Hz}$,	$\tau = 6$ ms.[11]
Our model gives:	$E = 43$ V/W; NEP = 0.7 nW/$\sqrt{Hz}$,	$\tau = 8.3$ ms.

We find that we can explain the sensitivity and the noise floor of the thermopile quite accurately with our simple model. We conclude that the thermal noise of the resistor is the primary noise source of the thermopile.

---

[10]http://www.hamamatsu.com/resources/pdf/ssd/t11262-01_kird1112e.pdf.

[11]http://www.heimannsensor.com/Datasheets/Datasheet-2b-TO18_rev2.pdf.

1.2.4 Sensor Electronics: Building the System

When we have a thermopile, it is time to think of the system. The following are the tasks of the sensor electronics in the case of a pyrometer:

1. Amplify the small voltage from the thermopile.
2. Compensate differences in sensitivity of the specific item.
3. Compensate for cross sensitivity to temperature.
4. Calculate object temperature.
5. Display object temperature and communicate it to an electronic port.

Do we really have to care for temperature cross sensitivity? The thermopile measures the temperature difference, so absolute temperature is not relevant. This is only partially true, since the thermoelectric coefficient of polysilicon and the wall heat transfer change with temperature, so the sensitivity will change as well. We have to compensate for this effect. To measure the system temperature, most thermopiles have a thermistor mounted at the side of the thermopile. To use this value for temperature compensation of the pyrometer, there are two strategies: the analog one and the digital one.

The first pyrometers had sensor electronics as shown in Figure 1.14. Thermopile and thermistor signals are amplified by two trimmable amplifiers, and their outputs are added. For calibration, the system is exposed to black surfaces with different temperatures. Then, the sensor is slightly heated and the measurements are repeated. This allows a calibration of the two amplifiers.

This approach works quite well for simple temperature measurement, when an accuracy of $\pm 1\,°C$ is sufficient. The system allows temperature compensation with one parameter, and in this way, a temperature cross-sensitivity, which is linear with temperature, can be compensated. However, the thermoelectric coefficient is a nonlinear function of temperature, and nonlinear effects cannot be compensated by one parameter only.

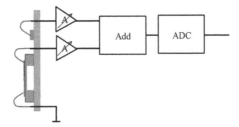

Figure 1.14   Analog temperature compensation.

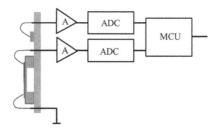

Figure 1.15 Digital temperature compensation.

An important application is the infrared ear thermometer. The temperature of the skin in the ear channel is measured, from which, the core temperature of the body can be calculated. German law for medical measurement prescribes an accuracy of ± 0.1 °C for body temperature measurement. This cannot be achieved by a linear approach, and we need to work digitally (Figure 1.15). We amplify both signals and convert them separately. Everything else is done in the digital world, where nonlinearity could be calculated in the micro controller unit. In most applications, this calculation is not done in the measurement system. Instead, all the calibration data, including sensitivity and temperature compensation, are calculated once and stored in the system as a three-dimensional look up table. All the MCU has to do is a linear interpolation in this table.

We learn two important lessons of general validity:

1. Measurement engineers are often faced with the following three obstacles:
    1. Noise
    2. Temperature cross-sensitivity
    3. Baseline drift
2. There is a trade-off between bandwidth and noise. We can make a fast system or a system with high resolution, but in many cases, we cannot get both at the same time.

1.2.5 The Alternatives: Pyroelectric Sensor and Bolometer

When we reconsider the effects for temperature measurement in Table 1.2, there is also the pyroelectric effect. It gives a signal when the temperature of the material is changed. In many applications, this is the case, and in the sensor market, we find a competition between thermopiles and pyroelectric sensors.

To build the pyroelectric sensor, we do not need a membrane on a silicon chip. We just use a small piece of crystal such as lithium tantalate ($LiTaO_3$) to realize a small measurement spot. We apply a metal pad on it and contact it with a thin bond wire. To achieve thermal isolation, we mount the crystals on small plastic pillars. As electric output signal, the crystal will give us a voltage, caused by a very small charge on the surface. This means that we have to measure close by with very high entrance impedance, which can be done using an op-amp directly at the side of the sensor, which is used as a charge amplifier. Also here, we have to care for temperature compensation. When the system is at high temperature, the response will be different even for identical IR power input. To compensate this, we apply a second identical crystal 2, connected antiparallel to crystal 1. While crystal 1 is covered with an IR-absorbing layer, crystal 2 is plated with gold to reject the IR power.

To be complete, we have also to regard the last method for temperature measurement, thermoresistivity. Just put a thin film resistor on the membrane, and you get a bolometer. The big disadvantage of bolometers is the fact that they measure the absolute temperature of the membrane. Now, when I use two in a Wheatstone bridge to compensate for temperature, I measure a small difference of two large numbers, such as 300.02 K – 300.01 K. Even for zero radiation, you do not get two zero values, but two large values with, hopefully, vanishing difference. This is sneaky and temperature problems arise always in the use of bolometers. We learn a general rule of making sensors: when offsets, drifts, and nonlinearities can be eliminated directly at the sensor on the hardware side, then do so. This is always better than the electronic compensation later.

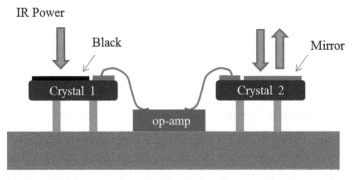

Figure 1.16 Pyroelectric sensor for thermal radiation. Two pyroelectric crystals are placed, one to sense and the other for temperature compensation.

On the other side, bolometer technology is simple, with only one metal layer. And noise can be low: when I apply a small resistor, thermal noise becomes small, too. De facto, bolometers are just used in special applications when extremely low noise is the aim to achieve.

1.3 A Topical Application Example: Infrared Gas Sensing of Ethylene

Of the food that is produced on this earth, two-thirds is consumed and one-third is lost. A large loss happens during transport, due to wrong cooling, mold, wrong humidity, insects, etc. The supervision of food transport by sensor nets is a fascinating and an important research topic. One important parameter for fruit transport is ethylene gas (C_2H_4). Ethylene triggers ripening of banana and many other fruits. When a banana ripens, it emits ethylene; when the banana senses ethylene, it starts ripening. Hence, one ripening fruit can trigger others to ripen. When we can measure ethylene in a fruit container, we can ask the fruits themselves what their status is. We need to measure low concentrations, and we need high selectivity versus other gases, which may be around.

Ethylene absorbs infrared light at 9 µm due to a vibration of the atoms in the ethylene molecule. Therefore, we can use infrared sensors (thermopiles or pyroelectric sensors) to measure it. Figure 1.17 shows the measurement system. IR is emitted from a thermal light source, such as a hot wire or a hot membrane. It passes through a cuvette filled with the air containing traces of ethylene. At the end of the optical path, there is a small band optical bandpass filter for 9 µm wavelength and a thermopile detector. When there is ethylene

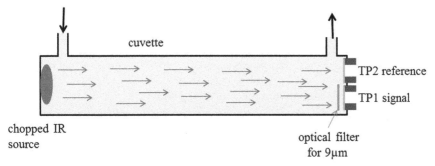

Figure 1.17 Non-dispersive infrared spectrometer (NDIR) for gas analysis. The absorption of 9 µm IR light by ethylene molecules is used to measure the concentration of ethylene in air.

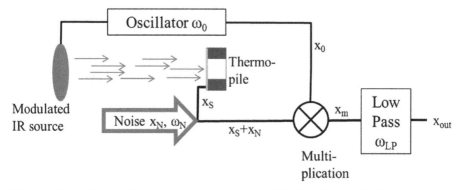

Figure 1.18   Reducing the noise by synchronous demodulation using the lock-in principle. Only a signal at the oscillator frequency ω_0 can pass, and all other signals are suppressed.

in the cuvette, the sensor signal is attenuated. The figure shows a two-beam spectrometer. The second thermopile measures broadband infrared. It is used to detect and compensate drifts in the intensity of the light source.

We have to measure ethylene in very low concentration. To achieve a resolution in the ppm (parts per million) range, we apply a trick: we chop the light source at 10 Hz. Then, we analyze only the part of the sensor signal that is modulated at 10 Hz $\pm$ 0.5 Hz and neglect everything else. In this way, we get rid of almost all thermal noise. This measurement trick is called lock-in amplification or synchronous demodulation; it is our most powerful tool to reduce noise.

Figure 1.18 shows the electronics of the lock-in system. It has an oscillator to modulate the light source with ω_0. The same oscillator is then multiplied with the signal of the thermopile. This product is finally lowpass filtered with a cutoff frequency ω_{LP}. In the case of NDIR, $\omega_0 = 10$ Hz and $\omega_{LP} = 0.5$ Hz. How is this electronics working?

The light source is modulated by the oscillator and it emits

$$x_0 = \cos \omega_0\, t. \tag{1.40}$$

The sensor receives this signal, and the sensor signal is:

$$x_S = A \cos \omega_0\, t \tag{1.41}$$

We recapitulate from the math course that:

$$\cos \alpha \cos\beta = 1/2 \cos(\alpha + \beta\,) + 1/2 \cos(\alpha - \beta)\,. \tag{1.42}$$

After multiplication, we get:

$$x_M = x_0 x_S = A \cos^2 \omega_0 t = \frac{1}{2} A \cos(2\omega_0 t) + \frac{1}{2} A \qquad (1.43)$$

The cosine term is fast oscillating and will be removed by the lowpass filter. The constant term stays and allows us to measure the amplitude A.

Now we consider noise, which is superimposed over the sensor signal as indicated by the big red arrow in Figure 1.18. Noise is chaotic, but we can, in any case, consider it to be composed of many Fourier components with different frequencies. For the Fourier component with frequency ω_N,

$$x_N = A_N \cos \omega_N t$$

the product is:

$$
\begin{aligned}
x_M &= x_0 x_N \\
&= A_N \cos \omega_0 t \, \cos \omega_N t \\
&= \frac{1}{2} A_N \cos(\omega_0 - \omega_N) t + \frac{1}{2} A_N \cos(\omega_0 + \omega_N) t \qquad (1.44)
\end{aligned}
$$

Case A: The noise frequency is not close to the signal frequency:

$$|\omega_N - \omega_0| > \omega_{LP} \qquad (1.45)$$

Then, both terms are oscillating fast and will be suppressed by the lowpass filter. The noise will be totally removed.
Case B: The noise frequency is close to the signal:

$$|\omega_N - \omega_0| \leq \omega_{LP} \qquad (1.46)$$

In this case, the second term ($\omega_0 + \omega_N$) is oscillating fast and will be removed by the lowpass filter, but the first term ($\omega_0 - \omega_N$) is oscillating slow enough and will stay.

So, we see that the lock-in system removes most of the noise, and only a small component very close to the signal frequency remains. The synchronous multiplication realizes a small bandpass around ω_0. Why then do we not just use a bandpass? The reason is in electronics: a bandpass filter is much more difficult to construct than a lowpass filter. This type of noise reduction is one of the most powerful tools in sensor electronics.

We find again the trade-off between bandwidth and noise: We can reduce the noise best when we put the final lowpass at a low frequency. This is achieved at the cost of reduced bandwidth or, in other words, a slow system, as discussed in Figures 1.11 and 1.12.

We did not look at the phase yet. When you do the calculus again considering phase shifts, you will find that not only all components with wrong frequency are removed, but also all components with wrong phase, which even increases the noise reduction power of synchronous demodulation.

The idea of lock-in can be used whenever a measurement principle works with an active stimulation of a vibration. The strategy is to stimulate a vibration, use it to sense, and remove noise by synchronous demodulation. We will use this idea several times when we look for measurement methods with extremely low resolution. Examples are the angular rate sensor (micro gyro) and the fluxgate sensor for the magnetic field strength.

1.4 Important Terms and Definitions for Sensor Characterization

When we discuss sensor characterization, we have to define some important terms. We all have a feeling what nonlinearity is, but "nonlinearity better than 2%" – what does this really mean?

The characteristics of a sensor describes the output (volt, sometimes ohm) as a function of the input (physical quantity, e.g., force).

The sensitivity S (sometimes also R for responsivity or E for the German "Empfindlichkeit") is the ratio of the sensor signal to the physical quantity to be measured.

$$S = \text{electric output signal/physical input signal} \qquad (1.47)$$

The range or full scale is the maximum value a system is specified for. The allowed over-range is the maximum value, which it will not be destroyed by, but the characteristics in the over-range is not specified.

The resolution is the smallest signal that can be distinguished from a neighboring value. Limits of resolution may be:

– Noise
– Number of bits in A/D conversion
– Number of decimal places on the display

But what does this mean? "I can distinguish a value from a neighboring value?" This needs to be defined quantitatively. A way to do so is to calculate the standard deviation σ of the sensor data. We assume 3σ as resolution, since two values are clearly separable when they are 3σ apart. Take care

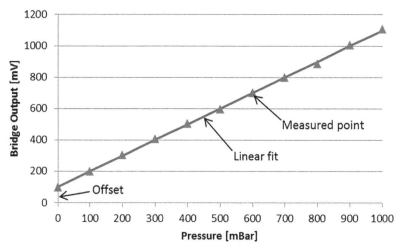

Figure 1.19 Characteristics of a pressure sensor.

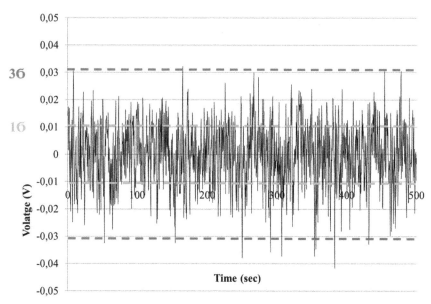

Figure 1.20 Noise of a sensor with error bands of 1σ and 3σ. In the 1σ band, we find 68.3% of all points; in the 3σ band, we find 99.7%. In most cases, 3σ is defined as the resolution.

when you read data sheets from sensors! Some people use 1σ as the measure for resolution, which makes their sensors look really good on the paper. Figure 1.20 shows a typical noise signal with error bands of 1σ and 3σ.

Accuracy is the maximum difference between the sensor output and the physically true value, which is guaranteed by the supplier under all specified circumstances (e.g., working temperature: $-40\,°C$ to $+120\,°C$). The resolution can be measured. The accuracy is specified by the supplier.

Example: A force sensor might have:

- A range of 1 N
- An allowed over-range of 2 N
- A noise equivalent force of 0.1 mN
- After a A/D conversion of 8 bit, a resolution of 4 mN
- An accuracy of 10 mN in the specified temperature range of $-20\,°C$ to $+80\,°C$

"Class 1" normally means maximum deviation is 1% of the full scale. Very often, the accuracy is defined for different regimes as a list or as a plot.

When we define a sensitivity S, we assume linear characteristics. Every sensor has a certain deviation from these ideal characteristics, its nonlinearity. How can the degree of nonlinearity in a sensor's characteristics be quantified? One method is "end point straight line": Draw a straight line over the full range, measure the maximum difference between the straight line and the sensor output, and divide by the span. In the example of Figure 1.21, we find a maximum difference of 25 V at a power of 5 W (arrows). The span is 160 V, so we have a nonlinearity of 16% (end point straight line). Of course,

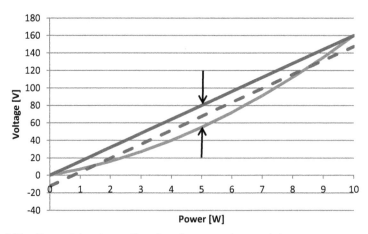

Figure 1.21   Determining the nonlinearity of a sensor characteristics.
Blue: Nonlinear characteristics
Red: End point straight line
Dashed red: Best-fit straight line.

this linear approximation is not the best one possible. We could use "best fit straight line", the dashed line 12.5 V below. Then, the maximum difference is 12.5 V and nonlinearity calculates as 7.8% (best fit straight line). The price we pay is an artificial offset: the linear fit will not pass the origin any more. Please note that nonlinearity depends strongly on the measurement range you define. In Figure 1.21, when you reduce the range to $0\ldots4$ W, nonlinearity (end point straight line) reduces to 10%. Again, take care when you read data sheets! How did they define nonlinearity?

1.5 Excursus to Wireless Sensor Networks

We want to get temperature and other measurement data from many places, but we do not want to deploy kilometers of copper wire. Then, a wireless sensor net is the solution. At this moment, wireless sensor nets are a big topic in research and technology for electronics and for informatics. There are two strategies to make a sensor wireless:

- We can transmit data and energy by the electromagnetic field using RFID technology. This is an established technology, but it is restricted to a range of few meters.
- We use a sensor node that is powered by a battery and communicates by radio. A number of those make a wireless sensor net.

By deploying many small sensors, we can supervise large areas. The single sensor node will not be able to communicate over long distance, but we can apply multi hop and send from one node to the next. Sensing can be highly parallel to assure coverage. Today, commercial sensor nodes have a size of 5 cm (Figure 1.22), while in research, nodes as small as 5 mm are demonstrated[12].

It turns out that sensor nodes should not just sense and communicate, but local intelligence and decision making are needed for the following two reasons:

- Energy: Electronics needs very little energy, sensing needs little, but communication needs a lot. Therefore, we use local intelligence to decide what needs to be transmitted, and the motto is: reduce communication to improve communication.
- Robustness is a big advantage of wireless sensor nets. When one sensor is destroyed, others can replace it. Neighboring nodes increase their

---

[12]http://www.happonomy.org/get-inspired/smart-dust.html.

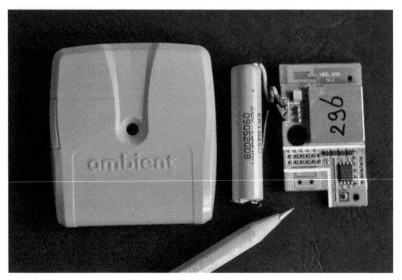

Figure 1.22 Commercial sensor node for temperature and humidity measurement by ambient systems.[13] The sensor node transmits at 2.4 GHz. With one battery, it can work for about 6 months.

activity, and new communication paths are established. When decisions are made locally by simple algorithms, the whole system becomes much more robust. And the system becomes more resilient, i.e., it gains the ability to recover when it is injured.

As an example for a wireless sensor net, I would like to discuss the "intelligent container"[14], a sensor net used to supervise the transport control of perishable goods such as fresh food[15]. Today, of all food grown on earth, only two-thirds is consumed, and one-third is lost in transport or storage. The idea of the intelligent container is to reduce wastage of food using sensors. What do we measure to know how transported fruit are doing?

---

[13]https://www.berlinger.com/de/temperaturueberwachung/produkte-hardware/produktueb ersicht/

[14]http://www.intelligentcontainer.com/

[15]Jedermann, R., C. Behrens, D. Westphal, W. Lang: Applying autonomous sensor systems in logistics combining Sensor Networks, RFIDs and Software Agents. Sensors and Actuators A 132(1), 2006.
Lang, W., Jedermann, R.: What can MEMS do for Logistics of Food? Intelligent Container Technologies—a Review. IEEE Sensors Journal Vol. 16, No. 18, pp. 6810–6818, 2016.

– Temperature is the most important quantity in fruit transport. Every cooled reefer has two or three thermometers, but it turned out that temperature gradients within a container can be quite large, and more measurement points are needed.
– Humidity: when it is too dry, fruits dry out, but when it is too humid, then fungus may grow. When there are temperature gradients in the container, there will also be gradients in relative humidity.
– Acceleration: Imagine a container with apples, dropped 0.5 m from the crane hook to the concrete floor at unloading. At first, you do not see anything, but next day, the apples develop brown marks and you cannot sell them anymore. For this reason, acceleration should always be monitored.
– So far, we measure the transport conditions around the fruit. Can we also measure the fruit itself? Is there a way to ask a banana how it is doing? In fact, there is! Bananas communicate by emission of ethylene. When under stress or when ripening, they emit ethylene. On the other hand, ethylene triggers ripening; hence, one ripe fruit may trigger other fruits around to ripen faster. Ethylene may be measured by NDIR (non-dispersive infrared) spectroscopy as discussed, or it may be measured with miniaturized gas chromatography systems.
– Biological threads: Is there fungus in the fruit? Are there insects in the container?

The next step is local intelligence. Where are calculations made, and where are decisions taken – locally in the container or somewhere in a logistic company on land? The container is a good example for how local decisions are advantageous.

Communication from the container to the main office is expensive. We do not want to send temperature data every 10 minutes. We want to measure every 10 minutes and then we decide whether the temperature is appropriate or not. If it is appropriate, one message per day is sufficient, and if not, we will send more details.

We need information about the status of the fruit. Is the banana still green or already ripening? Are the apples still ok for 2 more weeks? We need decision support tools such as the shelf-life estimator, an algorithm that predicts expected remaining shelf-life based on sensor data and on a biological ripening model, as shown in Figure 1.23. The shelf-life model predicts how biochemical reactions are accelerated by a temperature increase.

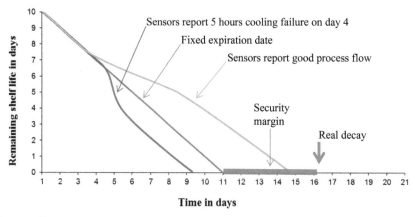

Figure 1.23   Remaining shelf-life estimator—the most important decision support algorithm in food logistics.

I assume that my apples will be good for sale for 10 more days when the storage temperature is 3 °C. I set a fixed "best consume before" or "eat by" date on day 11. Now imagine due to a power failure, temperature increases to 5 °C for 10 h on day 4. Then, the remaining shelf-life estimator will reduce the shelf-life accordingly, as shown in the figure.

The "eat by date" is chosen with a security margin. The real decay may be sometimes between days 15 and 18. Using sensors, I can see that the apples are doing fine and I can also prolong shelf-life as shown by the green line in the figure. In reality, this type of security margin is the source of enormous and avoidable food waste. In a future scenario, I would still sell and eat these apples on day 14, because due to sensor data, I know that they are still good.

When the sensor net tells me that there is problem on a ship on sea, what action can I take? There are two example scenarios:

- The FEFO scenario: Today, the logistic scenario is FIFO: first in, first out. For food, we would better apply a new scenario: FEFO: first expire, first out. When we know that a certain container has problems, we can change the allocation so that this fruit is consumed as soon as possible. Logistic modeling calculations show that shelf–life-based stock rotation might reduce loss considerably[16].
- Early warning: Even when we cannot save the fruit, real-time information can prevent large losses. Imagine some quality problem in banana

[16]Koutsoumani K, Taoukis PS, Nychas GJE: Development of a safety monitoring and assurance system for chilled food products. International Journal of Food Microbiology **100** (1–3), pp. 253–260, 2005.

plantation. The fruits are shipped, and they start ripening in the container before they should; a hot spot is developing and spreads out. When the ship reaches Europe after 3 weeks, the whole container is lost, and even worse, two other ships are already on sea with the same quality problem. The intelligent container network would see the problem beginning one week after the ship left port; it will give alarm and action can be taken on the plantation. In this way, a large loss is prevented.

Questions

- What effects do you know for temperature sensing?
- Explain the thermoelectric effect.
- Explain the pyroelectric effect.
- How does a thermopile work?
- Write down Nyquist's equation for thermal noise. Explain the terms. What can we learn for noise reduction from this formula?
- Estimate the noise of a thermopile.
- What material do we use for the thermopile membrane? Why?
- What do we use as thermocouple material in the thermopile? Why?
- What is the difference between accuracy and resolution?
- What is a time behavior of first order?
- Write down the step response of a first-order time behavior system. Draw response versus time.
- Name three distortions of the sensor output caused by slow response.
- Explain the pyroelectric effect.
- How many measurement points are within a band of $\pm 3\sigma$ around the average value?

2

Sensor Technology

2.1 Basics of Microtechnology

This is a book about sensing and not about sensor technology. There are dozens of interesting methods to make thin films, but I do not wish to explain all of them. There are excellent books about microtechnology, such as the book by Sami Franssila[1] or by H. Gatzen et al.[2], just to name two out of the many in store. My aim is to teach you the principles and the most important methods. Once you know how to ask the right questions, you easily can look up other methods.

When we consider making a structure like a thermopile, what type of technology could we use? We need membranes from thin insulating films, and we need electric interconnects integrated on them. The structures must not forcibly be very small, but there is no need for large size as well. We want to fabricate large numbers for a small price. All this points us towards the methods of microelectronics fabricated on silicon wafers:

- Silicon substrates with high surface quality are available.
- Several hundreds of devices can be fabricated on one wafer, which reduces the cost per device.
- The standard processes for thin-film deposition are available.
- The machinery for processes is available.
- Photolithography is an established technology for making structures in the µm range.
- Silicon technology in principle allows the integration of sensors and electronics.

---

[1] Sami Franssila: Introduction to microfabrication, Wiley.

[2] H. Gatzen, V. Saile, J. Leuthold: Micro- and Nanofabrication: Toole and Processes, Springer.

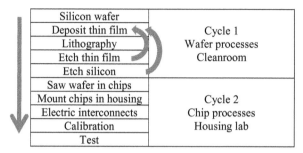

Silicon wafer	
Deposit thin film	Cycle 1
Lithography	Wafer processes
Etch thin film	Cleanroom
Etch silicon	
Saw wafer in chips	
Mount chips in housing	Cycle 2
Electric interconnects	Chip processes
Calibration	Housing lab
Test	

Figure 2.1   Basic steps of a micromachining process.

But there are disadvantages as well:

- We need a clean room, which means high initial investment.
- Excellent equipment is available, but very expensive.
- The high scale factor in production cost: When we make more than 100.000, a single sensor may be cheap. But making 100 or 1000 sensors require the setup cost and thus a single sensor will end up being very expensive.

The basic idea of making a micro sensor is shown in Figure 2.1 at the example of a thermopile.

We start with a silicon wafer. Then we deposit a thin film. We cover it with resist, and we make structures in the resist using optical lithography. Next, we etch the film, where it is not covered by resist, and in this way, we transfer the pattern from the resist in the film. This process is repeated, possibly several times. For the thermocouple, we deposit and structure polysilicon, and then aluminum. Until now, this has been the standard procedure of microelectronic technology. Then the desired membrane is prepared by etching the bulk of the silicon, which is a new process. When etching thin films, we remove 500 nm, but for setting free the membrane, we remove the full thickness of the wafer, 500 μm. Once again, we return to deposition to make and structure the final insulation layer for passivation. Now the wafer processes (cycle 1) are completed, and the wafer leaves the clean room.

In the housing lab, we first saw the wafer into many small chips. The chip is glued on a socket; thin wires are bonded for electric contact. We calibrate the sensor and finally test it. In the housing lab (cycle 2), every process is done at the chip level. Hence, we try to perform as many operations as possible at the wafer level in cycle 1. It is more effective and cheaper to process or test one wafer than to test 1000 sensor chips separately.

This chapter focuses on cycle 1: making thin films, structuring them by photolithography and etching, and structuring silicon. Housing technologies are discussed in Chapter 3 taking the instance of a pressure sensor.

2.2 Thin Films

2.2.1 Silicon as a Substrate

Our substrates are silicon wafers. Sometimes, we really require semi-conductivity, e.g. when we make pressure sensors. For the thermopile, we would not forcibly need silicon, and do not use semi-conductivity. We use silicon wafers for two other reasons. First, the quality of the surface is very good; it is polished down to 3-nm roughness. This is needed for thin-film technology. The roughness of the substrate must be much smaller than the film thickness; otherwise, a film might be interrupted somewhere. Second, silicon wafers are the standard material for the process machines from microelectronics. Every sputtering or lithography machines has holders for silicon wafers, while for other substrates we would have to construct extra equipment.

Silicon wafers are single crystal material with high purity. For sensors, we apply 100-mm-diameter (4 inch) or 150-mm-diameter (6 inch) wafers. For microsystems, wafers of size >150 mm normally do not pay out because the lot sizes are not so high. Microelectronic industry uses 300-mm (12 inch) wafers, and 450-mm technology is coming up. Single crystalline silicon is grown from the melt in the form of large cylindrical crystals; these are cut in wafers, ground and polished. For electronics, the front side is polished; for microsystems, wafers are polished on both sides to be able to apply lithography on the back as well. A large variety of doping is available. Thickness rises with diameter; the typical values are 525 µm for a 100-mm wafer and 700 µm for a 150-mm wafer. Thickness variation is critical when making membranes; a typical value is a TTV (total thickness variation) of ±15 µm.

2.2.2 Thin Films

Thin films have a thickness of 20 nm to 2 µm. Dust particles in the surrounding air are much larger than these. A particle of size 10 µm in a film would interrupt the film, and, if placed in an interconnect which is 5-µm wide, it would interrupt this electric line and destroy the function of the chip. Hence, we have to make thin-film technology in a clean room. An appropriate surrounding would be clean room class 5 according to ISO 14644-1, which means that there are less than 3520 particles > 0.5 μm in a cubic meter of air (or 100 particles in a cubic foot).

Applications of thin films are many:

- Microelectronics, such as metallization and insulation in integrated circuits.
- Sensors, such as thermopiles and strain gauges.
- Optics, such as antireflection coatings.
- Decoration like gold plating.
- Anticorrosion like chromium plating.
- Tribology, e.g. SiC hard coatings on a tool.
- Packaging of food: In a tetra pack, the paper of the package is coated with thin-film aluminum and polyethylene to prevent soaking of the paper and to preserve the liquid content.

When we compare a thin film of 200-nm gold and a gold wire of 1-mm diameter, then we find that there are differences in the behavior of the material. The typical properties of thin films are:

Microstructure

The microstructure is different from the bulk. Thin films can be amorphous, but they can also have many different structures like columnar or crystalline. The amount of grain boundaries can change mechanic strength and electric resistivity. Some morphologies like columnar structure show in thin films, but never in a bulk piece of the same material. The way a film grows has to do with the way it is deposited, the most important factor being energy and temperature. This will be discussed in the next chapter.

Voids

Thin films have voids. Sometimes, a spot is left out during deposition. Then it can happen that the deposition continues and overgrows this spot, and a void is generated. A void may also go through the whole film, and then we call it a pinhole. The materials we use for deposition are very pure, and so we should not expect foreign pollutions, but we use process gases such as argon and hydrogen. These can be incorporated into the film and contaminate it. This implies that in our gold film not all volume is filled with gold, and therefore, the density can be a little smaller than for the bulk material:

$$\text{Density: } \rho_{\text{Thin film}} < \rho_{\text{Bulk}}$$

Resistivity and TCR

Electric resistivity is larger than that for the bulk. This is a consequence of the smaller density, but there is another important reason: electric resistivity means that travelling electrons loose energy when they collide with some obstacle. This may be a grain boundary, or it may be a thermal excitation in the film, called a phonon. In thin films, there is one more mechanism of collision: an electron can also loose energy by colliding with the surface. This makes a higher rate of collisions and therefore a higher electric resistivity.

$$\text{Electric resistance: } \rho_{\text{Thin film}} > \rho_{\text{Bulk}}$$

The temperature coefficient of resistivity (TCR) of the thin film generally will be lower than for the bulk. This can be understood from the same argument. When temperature rises, then there are more phonons. The rate of collisions rises, and so does resistance. Hence, metals generally have a positive TCR. The number of collisions with the surface does not rise with temperature. The same holds for collisions with grain boundaries, voids and many other kinds of film imperfections. Hence, the additional mechanisms of resistivity, which thin films show on top of the bulk material, do not show a TCR. In this way, the overall TCR of the thin film will be lower than for that of bulk.

$$\text{TCR}_{\text{Thin film}} < \text{TCR}_{\text{Bulk}}$$

Both density and specific resistance are difficult to measure because we have to know the exact thickness to calculate them. The temperature coefficient is easy to measure, since it is a relative measurement. We just measure any resistance, then put the wafer on a hot plate and measure again.

Stress

The most fascinating and probably the most dangerous of all thin-film specialties is the internal stress. We know film stress from daily life experience: When paint dries, it shrinks, there is less film than area, and there is a tensile stress. When we have bad luck, the stress is larger than the paint can sustain and there will be cracks. On the other hand, when an apple dries out, the volume shrinks while its skin does not. There is more skin than flesh underneath; this will make compressive stress and there will be wrinkles.

When a film on a wafer exerts stress, the wafer as a whole will bend a little bit, as shown in Figure 2.2. For compressive stress, it will be buckling; for tensile stress, this will be a bow like deformation. This actually happens on silicon wafers, the bow will not be much, maybe 50 μm, but it will be a problem since we cannot focus optical lithography any more. This is also how

| Compressive stress in thin film causes membranes to buckle | Tensile stress causes cracks, delamination of films, braking of membranes |

Figure 2.2   Thin-film stress will deform a wafer to a buckle or a bowl.

we measure film stress: We measure the surface profile of a wafer, then we deposit a film, then we measure again. Stress in films can destroy our work in a number of ways: High tensile stress can deform microstructures and it can cause delamination of films. Compressive stress can make membranes buckle.

Where does the stress come from?

- Thermal stress is obviously seen. When we deposit film at high temperature and then cool down, then there will be thermomechanical stress, since silicon and the film have different coefficients of thermal expansion.
- When foreign atoms are caught in the film in large numbers, they can cause compressive stress. An example is silicon nitride deposited by low temperature plasma CVD (explained below), which has a lot of hydrogen incorporated.
- The last type is stress caused by the growth process. This intrinsic stress is difficult to understand and hard to predict. Some reasons are as follows:

  o Bonds are either too long or too short. Imagine a human chain as sometimes done in political demonstrations. A person standing with open arms will cover a length of around 125 cm. When there are 80 persons on 100 meters, the human chain has a perfect structure. When there are 200, the chain is compressed, it will be wavy or wrinkled. When there are only 60 people, they have to stretch and there will be tension. The same happens in thin films. For example, in silicon nitride, there is an optimum distance between silicon and nitrogen. When the deposition process puts them at larger distance, there will be tensile strain. Actually, a very important deposition process for nitride (LPCVD) generates extreme tensile stress. But why do the particles not deposit with the distances they would need to be stress free? This is not obvious, and intrinsic stress is hard to predict. We have to measure it for each new deposition process we develop.

- ○ Crystallites may fight for room: Initially, there is a nucleus with size of some atoms. When the film growth proceeds, then the crystallite attracts more atoms which deposit on its surface and the crystallite grows. But there is a neighboring crystallite, which also attracts atoms. And in this way, more atoms are drawn between the crystallites than there is space, and compression is generated.
- ○ A small void separates two surfaces. When they are close enough to each other, then there will be surface attraction, and the void becomes a source of tensile stress.

2.2.3 Thin-Film Deposition Methods: An Overview

All the methods to grow thin films have been classified into several families:

A. Physical vapor deposition (PVD) uses physical methods (no chemical reactions). It is mainly used for metal films made from single atoms (gold, aluminum, etc.).
B. Chemical vapor deposition (CVD) uses a chemical reaction to form a compound which is then deposited. It is mainly used for insulating films such as silicon oxide and silicon nitride. It is also used for thin-film deposition of silicon (Poly = polycrystalline silicon).
C. Liquid deposition and drying. The most important application is spin-on of a liquid resist, which is dried to become a solid film.
D. Electroplating: Ions in a solution are transported to a surface by an electric current and incorporated in the surface but electrochemical processes. This is used to obtain metal layers of thickness greater than 1 μm.

2.2.4 Physical Vapor Deposition (PVD)

The most simple method of PVD is <u>evaporation</u>. A metal is heated in high vacuum by electric current or by an electron beam (e-gun) until it melts and evaporates. The atoms sublimate onto a cooler substrate and generate a thin film. This type of PVD is usually used for laboratory experiments, exotic materials and sometimes for metallization.

When the particles touch the cold surface of the substrates, they adhere where they hit. The kinetic energy the particles gain due to evaporation is small. They cannot travel on the surface after adsorption to find a good lattice site. Due to the low surface mobility, the particles cannot fill all voids and the resulting film will have a poor quality.

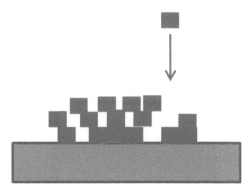

Figure 2.3 Film growth at low energy.
Adsorbed atoms stick where they touch. Gaps are not filled; shading occurs. The result is voids, pinholes, and bad surface quality.

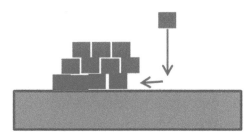

Figure 2.4 Film growth at high energy.
The atoms have kinetic energy. After adsorbing, they can wander around until they find a good lattice place with many bonds to neighbors. Gaps are filled. The result is good film quality.

Figures 2.3 and 2.4 show this kind of Tetris game for atoms. When there is low mobility, then atoms stick where they hit. Gaps cannot be filled and voids remain. The result is bad film quality. When energy increases at higher temperature, the particles diffuse around on the surface until they find a good place. What is a good place? This is a position where the particle can make many bonds with neighbors. Making bonds reduces energy; every particle moves down on the energy scale, if it can. In this way, good mobility allows a film with more connections. As you can see in Figure 2.3, for low energy the particles have two to three direct neighbors, while for the high energy case in Figure 2.4 they have six neighbors. In the high-energy case, when shading might occur, the next particle will diffuse around; it will find this place, adhere exactly there and fill the gap.

The effect of low mobility on the growth of a regular crystal pattern can be observed in a lecture hall. Before the lecture starts, mobility is high. When

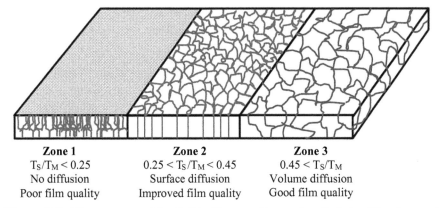

Zone 1	Zone 2	Zone 3
$T_S/T_M < 0.25$	$0.25 < T_S/T_M < 0.45$	$0.45 < T_S/T_M$
No diffusion	Surface diffusion	Volume diffusion
Poor film quality	Improved film quality	Good film quality

Figure 2.5 Three-zone model of thin-film growth as described by Movchan and Demchishin.[3] Figure adapted from Movchan and Demchishin.

someone enters the room and sees an empty place, he or she will ask people to let him go there. Voids are filled up and every seat can be used. When the lecture starts, mobility is inhibited. Latecomers will see empty places in the middle of a row, but since the lecture is already going on, they will consider it to be impolite to make people stand up to let them pass. In this way, latecomers will take seat at the side or in the last row. Sometimes they even stand, in spite of the fact that some seats remain empty in second row. This corresponds to bad crystal quality.

Obviously, the temperature of the substrate is important for film growth. A general morphology model was devised by Movchan and Demchishin (Figure 2.5). It shows three zones for film morphology. The zones form depending on the relative temperature of the substrate T_S in proportion to the melting temperature of the thin-film material T_M.

Zone 1: $T_S/T_M < 0.25$:
T_S = Temperature of the substrate, T_M = Melting point of the film material. Here the rule is mainly: particles stick, where they hit. There is no energy, and therefore no surface mobility. There is dendritic growth and shadowing. There are voids, bad adhesion to the substrate, and porosity.

Zone 2: $0.25 < T_S/T_M < 0.45$:
This temperature will allow surface mobility. Voids fill up. Each crystallite will grow vertically, but lateral growing together and merging of the crystals does not occur, since the energy is not sufficient for volume diffusion. In this

[3]B. A. Movchan, A. V. Demchishin: Phys. Met. Metallogr. 28 (1969) 83.

way, zone 2 films often show a columnar structure as shown in the figure. Adhesion will be much better than in zone 1, but still not really good.

Zone 3: $0.45 < T_S/T_M$:

Now not only surface diffusion occurs, but also volume diffusion. A particle can move around inside the already grown film and maybe find a better place. Recrystallization occurs; the columns can merge and make new crystallites. The columnar structure of zone 2 is replaced by a polycrystalline structure. We will find good film quality and good adhesion.

We want zone 3 films, of course. A problem of the PVD method of evaporation is that zone 3 is not achievable for high melting metals. For aluminum, the melting point is $T_M = 940$ K, and zone 3 starts at $0.45 \cdot T_M = 420$ K = 150°C. This will work, and evaporation can be used to make aluminum thin films. Can we make a good platinum thin film by evaporation? $T_M = 2030$ K, zone 3 starts at $0.45 \cdot T_M = 913$ K = 640°C. This will not really work. Heating up the substrate so much would generate problems such as thermal stress when cooling down.

To improve the film quality, a higher surface mobility is needed, but heating the substrate is not the solution. Therefore, a different way has to be found to supply the adsorbing atoms with kinetic energy. This can be accomplished using an electric field in plasma realized in a new PVD method called sputtering.

Let us remember the most important facts about plasma. In a sputtering reactor as shown in Figure 2.6, low-pressure argon gas is exposed to an electric field. Some argon atoms will be ionized, and these are accelerated by the field. They gain energy. When they collide with other argon atoms, they can ionize them. In this way, we generate a steady flow of Ar^+ ions accelerated to high energy in the reactor.

Now we want to use this energy for depositing gold. The gold comes from the target; this is a plate of solid gold that is opposed to the substrate. The electric field is generated by one electrode below the substrate; the second electrode is the target. This way, we generate argon ions (green in the figure), and we accelerate them versus the target. Finally, they hit the target, and they have quite some kinetic energy. What happens? The argon ions knock out gold particles from the target. These gold particles now travel through the gas, and they also have high energy due to the mechanic impact of the argon ions. This process of removing particles by hitting a surface with accelerated ions is called "sputtering off". The gold particles travel through the reactor and finally hit the substrate, where they are adsorbed. To allow them travelling,

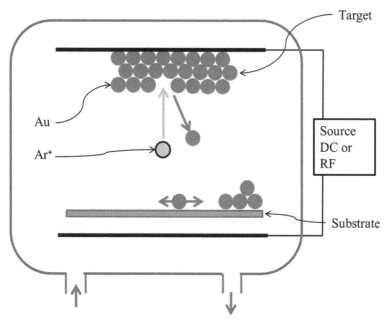

Figure 2.6 Sputtering: In a plasma, Ar atoms are ionized and accelerated towards the target. They knock out atoms which travel through the recipient and finally deposit on the target.

we need a mean free path which is about as long as the reactor is large, and so we need low pressure, about 0.1 Pa. We cannot go down to ultrahigh vacuum, since we need the argon to make the plasma burn. When gold atoms hit the surface, they have a high kinetic energy (much higher than the atoms in the evaporation process). Due to their high kinetic energy, the atoms have high mobility on the substrate surface and therefore move. When an atom comes to a place where it can bind with more than one other gold atom, it will favorably settle at this gold lattice side on the surface. The film has few voids and much better film quality than a film made by thermal evaporation.

The surface mobility of the deposited gold particles can be controlled by the power supply and the gas pressure. Therefore, we get a zone 3 film structure even for low substrate temperature. Sputtering can be done with DC current, but in most cases, we will use AC with a DC bias superimposed, which allows better process control. To increase the ionization, we have to care for many collisions between argon and argon, but we do not want to allow more pressure. For this, a magnetic field can be applied. This makes

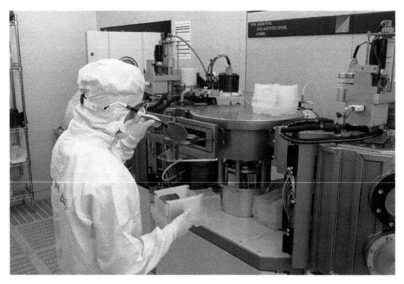

Figure 2.7   Loading a silicon wafer in a sputtering machine. The round vessel in the middle is the load lock chamber. At this moment, the door is open and a wafer is placed on the arm of the wafer handling system using tweezers. Right and left of the load lock chamber there are two process chambers for sputtering. You can see the cleanroom clothing in a class 5 cleanroom: overall, gloves, face mask and safety glasses.

the charged Ar^+ ions move in a spiral way through reactor allowing them a much longer path and more chances for interaction. This improved process is called magnetron sputtering.

The process looks simple, but machinery is difficult and expensive. We need a powerful vacuum system and a gas supply as well as a high voltage AC power source and the targets, which have to be water cooled to avoid overheating. The high voltage and the cooling water need a feed through into the vacuum recipient. After each deposition, we have to open the reactor to remove the wafer and load a new one, but when we open the vacuum recipient, humidity and other contamination will instantly cover the surfaces. To avoid this, most machines use a two-chamber system with a process and a load lock chamber. The load lock is opened, the wafer fed in, the chamber is closed and pumped down to a vacuum of 1 Pa (10^{-2} mbar). Then we open a second door between the load lock and the process chamber. We transfer the wafer with an automatic conveyor into the process chamber, close the door and pump again. The advantage such a system as shown in Figure 2.7 is that the process chamber is never contaminated with air.

When we can sputter off material from the target, we can also sputter off material from the wafer; this is called sputter etching. We do this to clean the wafer. In the thermopile process, after structuring polysilicon, we want to deposit aluminum. We load the wafer, and then we do sputter etching for a few minutes, which will remove about 10 nm of the material. Then we sputter the metal. The idea is to remove contamination and some thin oxide which might have been growing on the polysilicon in air.

Can we sputter an insulating film? Using glass as the target, a DC current could not flow, but AC excitation worked. Another way is reactive sputtering, when we sputter off silicon and allow oxygen in the reactor to make the silicon atoms oxidize while they fly from the target to the substrate. The process works, but is difficult to control. Most sputtering applications in microtechnology are metals, and for insulating layers, we have to look for other possibilities.

2.2.5 Chemical Vapor Deposition (CVD)

While metals are monoatomic, insulators are compounds of several different species: silicon-oxide, silicon-nitride and aluminum-nitride. We can make use of this fact for deposition, and this leads us to do chemical vapor deposition (CVD). In CVD, we deliver the two species by two different gases, which are called precursor gases. Then we initiate a chemical reaction. An example is silicon nitride. We feed the reactor with silane (SiH_4) and ammonia (NH_3). We apply energy to break the molecules so that they can recombine in a new form: silicon nitride Si_3N_4. The chemical reaction is as follows:

$$3SiH_4 + 4NH_3 + energy \rightarrow Si_3N_4 + 12H_2$$
$$\text{Silane} \quad \text{Ammonia} \qquad \text{Silicon nitride}$$

The first reaction product is silicon nitride; it adsorbs at the substrate and then forms a thin film. The second reaction product is hydrogen, which is a gas and we can pump it away. Silane is a dangerous process gas; when it comes in contact with oxygen, it instantly reacts, i.e., it burns at every concentration. It is nasty, but in silicon technology, you cannot avoid it.

A second possible precursor system is dichlorsilane ($SiCl_2H_2$) plus ammonia:

$$3SiCl_2H_2 + 4NH_3 + energy \rightarrow Si_3N_4 + 6HCl + 6H_2.$$

We can also deposit silicon oxide from silane and oxygen:

$$SiH_4 + O_2 \rightarrow SiO_2 + 2H_2.$$

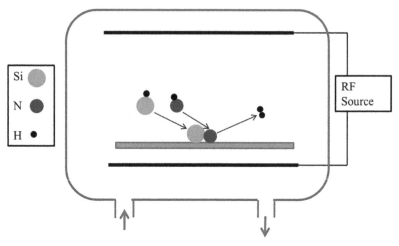

Figure 2.8   Plasma-enhanced CVD: The material is delivered as precursor molecules. The molecules are broken by the plasma energy; the parts recombine in a new film on the substrate. Example of silicon nitride Si_3N_4
Precursor 1: Silane SiH_4
Precursor 2: Ammonia NH_3
Chemical reaction: $3\ SiH_4 + 4\ NH_3 + energy \rightarrow Si_3N_4 + 12H_2$
The nitride is deposited, the hydrogen is pumped off by the vacuum pump
In the picture for simplicity only one atom of each kind is drawn in each molecule.

Another possibility for silicon oxide is the use of tetraethyl orthosilicate (TEOS = $C_8H_{20}O_4Si$) as a precursor.
CVD is also the best method to make polysilicon thin films from silane:

$$SiH_4 \rightarrow Si + 2H_2.$$

The question is now, how to bring in the energy and initiate the chemical reaction. You may guess that there are two possibilities again: temperature and plasma. A third possibility is laser CVD, but this is rarely used.

A reactor for <u>plasma-enhanced CVD</u> (Figure 2.8) has a gas pumping system and a radio frequency (RF, typically 13.56 MHz) source for driving the plasma. PECVD films are made at a pressure around 20 Pa to 200 Pa and at relatively low temperatures from 250 to 400°C. The energy of the plasma breaks the molecular bonds of the educts (silane and ammonia). Then the parts rearrange and form silicon nitride, which is deposited at the surface. The other reaction product (hydrogen) is in the gas form and can be pumped off. Due to the relatively low temperature, the surface mobility is low and the films have voids, pinholes and poor step coverage. A second problem is a large amount of hydrogen incorporated in the film, which causes large compressive stress. When the film is annealed, the hydrogen diffuses to the surface and desorbs, the stress is released. When we make a membrane from PECVD nitride, then it will buckle due to stress. After annealing, the membrane will be flat. Other important thin films fabricated with PECVD are silicon oxide and silicon carbide.

A better film quality can be achieved using high temperature. The process is called LPCVD (low-pressure CVD) since it applies not only high temperature, but also low pressure. The chemical reaction is the same as for PECVD, but the energy needed to initiate the reaction comes from the heat this time. LPCVD silicon nitride is an extremely good insulating material with great electric breakdown strength and also high mechanic strength. It is made at a temperature of 600°C and a pressure of 50 Pa. A 300-nm film can be used for a freestanding membrane of 3 x 3 mm. Those membranes are very stable; a pressure of 1 bar normally will not break them. Silicon nitride is also chemically stable and used as a masking layer for aggressive etchants. Yet there is one serious drawback: LPCVD nitride has an extremely high intrinsic tensile stress of 1.2 GPa. Steel would break under this load. It can deform small structures and it can peel off from the substrate. When you want to do LPCVD nitride on silicon, it is strongly recommended to apply an intermediate layer of silicon oxide. Oxide is weak and can relax stress, and it adheres well to both silicon and nitride. If you do not apply the oxide, you will risk the nitride peeling of the wafer by its own intrinsic stress. Si_3N_4 is stoichiometric oxide. Some people deliberately do apply nonstoichiometric oxide on the silicon-rich side, which has much less stress[4]. Other important LPCVD films are silicon oxide and polysilicon.

The process seems to be simple; we do not need an electric field. Yet the process is extremely difficult to handle because we must go to high temperature and must control this high temperature extremely well. A temperature of 605°C instead of 600°C would already make a noticeable change in the deposition rate. A small gradient over a 6-inch wafer would make inhomogeneous films. We use a three-zone oven as shown in Figure 2.9. The oven is a large tube made of glass. The oven has three electric ring heaters. When they are controlled well, there is a length of about 25 cm in the middle where temperature is very homogeneous. This zone is used for processing. There is space for many wafers, so we typically process up to 20 wafers in

---

[4]Thermophysical properties of low-residual stress, silicon-rich, LPCVD silicon nitride films. C. H. Mastrangelo, Yu-Chong Tai, R. S. Muller. Sensors and Actuators A, 23 (1990).

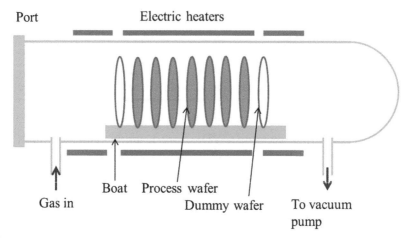

Port   Electric heaters

Gas in   Boat   Process wafer

Dummy wafer   To vacuum
pump

Figure 2.9   Three-zone furnace for high temperature processes. The furnace is a long tube made of quartz glass. The three heating zones guarantee a homogeneous temperature profile in the middle zone, where the process wafers are placed. Temperatures go as high as 1200°C. These ovens are used for oxidation, LPCVD, diffusion doping and annealing.

one process run. The wafers are placed in a glass holder called a boat. This is shifted into the oven with a long glass stick, and after some hours, it is taken and pulled out with this stick again. It reminds us of baking pizza, when you slid the raw pizza in the oven with a peel and, after 10 minutes, catch the hot pizza and draw it out. For LPCVD, the oven must have a vacuum system, a gas tight door and a gas inlet. Considering the size of some meters, the high temperature and the fact that all is made from glass, we understand why these ovens are expensive and vulnerable pieces of equipment.

When do I use PECVD and when LPCVD? The answer is: LPCVD whenever possible, it makes better films. We use PECVD when necessary because LPCVD is not possible due to the temperature. There is one more difference that is practically important: PECVD is a single side deposition, since the wafers lay flat on a substrate holder. In LPCVD, the wafers stand upright in the boat, and both sides are covered.

2.2.6 Oxidation of Silicon

Besides silicon nitride, the other important insulation layer of silicon technology is silicon oxide. It can be fabricated by CVD as discussed or by thermal oxidation of the silicon wafer. For thermal oxidation, the silicon wafer is heated to high temperatures (800°C to 1200°C) and exposed to air. The oxygen reacts with the silicon and an insulating layer of very good quality is formed. The oxygen molecules have to diffuse through the already-grown oxide film to the silicon interface to react with the silicon. This makes the process self-inhibiting. The film thickness does not raise proportional to time, but the process becomes slower with increasing film thickness.

To speed up, we must enhance diffusion of oxygen through the oxide film. Astonishingly, this works with water. The H_2O molecule is smaller than the O_2 molecule, and therefore will diffuse faster. When the wafer is not exposed to air but to hot steam, oxidation is much faster. "Wet" silicon oxide has very good adhesion to the silicon and good insulating properties. It is the preferred insulating film in semiconductor technology.

2.2.7 Surface Migration and Step Coverage

High surface mobility is not only essential to get films free of pin-holes, but also for step coverage. Imagine we deposit a layer 1 and structure it. When we deposit layer 2, the steps must be covered. Figure 2.10 shows the situation. If a point is on a flat surface, atoms from all directions above the surface will reach this point and be deposited. The angle of arrival is 180°. At the upper edge of the step, the angle is larger at 270°. In the step, the angle reduces to 90°. The number of arriving particles will vary for the different spots. We will find a certain film thickness at the flat parts, and an increase or lump at the upper edge. At the lower edge, we will find a recess (red flash), which may be a serious weak point of the film. Imagine the first film to be aluminum, and the second film silicon nitride. We might use the nitride to mask the aluminum against etching in a later step. If the recess is as bad as in Figure 2.11, then the etchant will penetrate the masking layer and destroy the aluminum. This type of insufficient step coverage is typical for PECVD films, which we use for final passivation of a sensor against humidity. Hence, passivation layers are quite thick, e.g., 2-µm passivation to cover 0.3-µm steps.

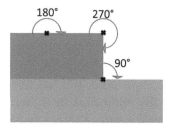

Figure 2.10   At steps and corners, the spatial angle of incident is changed.

Figure 2.11   Low surface mobility: in a recess, the film is very thin.

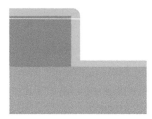

Figure 2.12   High surface mobility: the atoms travel, conformal thickness.

LPCVD, allowing large surface mobility, would allow a conformal film thickness as in Figure 2.12. The atoms will move around and cover everything. Unfortunately, we cannot apply it for the thermopile passivation, after doing aluminum metallization.

In microelectronics, the filling of small trenches is a crucial problem for technology. Here, LPCVD is the ideal method. First, it has high surface mobility. Second, it is done at low pressure and allows atoms to penetrate into deep trenches. Trenches filled with LPCVD are used as capacities for CMOS memory cells.

In order to revise all the methods of film deposition, a summary is given in Table 2.1.

2.3 Thin-Film Deposition Control

After each process step, we have to control the film. The first control is done by the experienced eye of a technician. Thin transparent films show interference colors, and with some experience, you see when the silicon nitride has the wrong thickness. Metallic films also show strange color or surface granulation when deposition went wrong. This is important, but not enough. We have to measure important film parameters directly after deposition. In transparent films, this is thickness. For metals, it is sheet resistance.

Table 2.1 Overview on thin-film deposition

	PVD		CVD	
Family	Physical Vapor Deposition		Chemical Vapor Deposition	
Materials	Layers made of single atoms, Metals		Chemical compounds: Insulators But also polysilicon	
Name of method	Evaporation	Sputtering	PECVD Plasma- enhanced	LPCVD Low pressure
Substrate temperature	Low T	Low T, but high energy	Low T	High T
Pros	+ Simple	+ Good film quality	+ Low T- Budget	+ Highest quality
Cons	− Poor film quality	− Effort − Target cost	− Poor Quality	− Temperature − Cost
Application	Exotic materials R&D	Metallization	Passivation	Insulation, membranes

2.3.1 Metal Films: Measuring Sheet Resistance

For metal films, we could measure thickness, but to do so, we would have to etch a step. The easier way is to measure electrical resistivity. We define the sheet resistance $R_\square$ ("R-square"), which is the specific resistivity ρ over the thickness D:

$$R_\square = \frac{\rho}{D} \tag{2.1}$$

Sheet resistance is the resistance one square of our film would have, independent of its size. It is always important to know the sheet resistance of our metallic layers. When it deviates, then either the film thickness is wrong or we have a major problem in the morphology or purity of the film.

Sheet resistance is measured without an etched structure using the four-lead probe shown in Figure 2.13. The film is touched gently with the four point contacts. The outer contacts are used to impress a measurement current I; the inner two are used for measuring the voltage V. This allows the measurement not being affected by contact resistance. The current lines will spread out in the thin film, but presuming that the film is thin with respect to the distance of the contacts, it can be shown that there is a strict relation between voltage and sheet resistance[5]:

$$R_\square = 4.53 \frac{U}{I} \tag{2.2}$$

---

[5]F. M. Smits: Measurement of sheet resistivities with the four-point probe, Bell System Technical Journal, vol. 34, 1958.

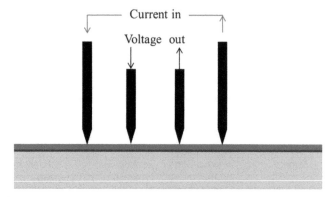

Figure 2.13   A four lead probe for the measurement of sheet resistance.

The advantage is that the measurement can be done directly after film deposition without having to etch for the making of test structures. A commercial four-lead probe has an automated xyz stage like a wafer prober and it can scan a wafer in a few minutes.

2.3.2 Dielectric Films: Interferometers

Transparent films such as silicon oxide or nitride are measured optically using interferometry. Let us reconsider interference of light at thin films (Figure 2.14).

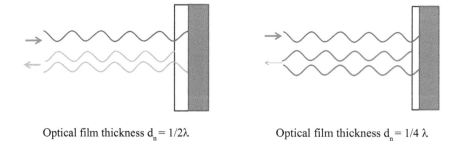

Optical film thickness $d_n = 1/2\lambda$
Optical path difference $2d_n = \lambda$
Constructive interference

Optical film thickness $d_n = 1/4\lambda$
Optical path difference $2d_n = 1/2\lambda$
Destructive interference

Figure 2.14   Optical interference of a wave with wavelength λ at a thin film with thickness d. The light enters into the medium with higher index twice, so there is a phase jump for both reflected waves.

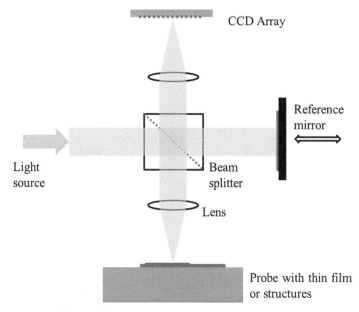

Figure 2.15 A scanning white light interferometer.

The analysis of the interference allows measurement of the optical thickness d_n of the film. This is correlated to the geometrical thickness d by $d_n = n \cdot d$. Hence, we need to know the index of refraction n of our insulating films. For thin-film measurement, we normally use white light interferometers. The illumination is done with white light, and we use a spectrometer for detection. In this way, we obtain the interference features for many colors simultaneously.

But interferometry can do more than measuring film thickness, it also can depict structures. Figure 2.15 shows a scanning interferometer. The heart of the device is a beam splitter that lets 50% of the light pass and reflects the other 50%. There will be two light beams arriving at the CCD array: one is reflected by the probe, the other at the reference mirror. These two can interfere constructively or destructively. To see the interference pattern, we can move the reference mirror back and forth a little bit. When the probe is scanned, the topography of the probe will modulate the optical path and generate an interference pattern. In this way, we obtain a 3-D picture of the probe surface. The scanning interferometer is an important tool to measure the step height. We use it to measure the thickness of metal films, but we have to etch a step in the film.

2.3.3 Doping

The electronic properties of silicon are defined by doping. Doping intentionally introduces impurities into a semiconductor for the purpose of modulating its electrical properties. Doping can be performed during the wafer production for the whole wafer or during processing as local doping at the surface. There are two important methods for doping during the process:

1. Diffusion: The wafer is heated in an oven in an atmosphere containing doping atoms. These atoms adsorb on the wafer surface and then, due to high temperature, diffuse into the silicon.
2. Implantation: Dopants are ionized and accelerated in an electric field to high kinetic energy. When the dopants hit the wafer surface, they penetrate into the material and dope it.

The dopants will only reach the surface layer (some hundred nm) of the material. Local doping can be performed when a part of the wafer is covered with a mask such as photoresist or oxide.

2.4 Wafer Bonding

Quite often, one wafer is not enough. Being able to join two wafers as a full wafer process and then continue technology with a sandwich wafer opens interesting possibilities:

- Close a cavity hermetically sealed.
- Get monocrystalline membranes with precisely defined thickness and electric isolation to the ground wafer.
- Join one wafer with CMOS electronics and another one with sensors.

We could do this during housing as a single chip process, but it is much wiser to do it at the wafer level, for all chips on a wafer simultaneously. Hence, we look for processes of full wafer bonding.

2.4.1 Silicon Fusion Bonding

One silicon wafer is bonded on a second one with an oxide layer in between. The lower one is called the handle wafer, while the upper one is called the device wafer. The main process steps are as follows:

- Deposit silicon oxide on the handle wafer.
- Bring both wafers in touch. Surface adhesion forces will connect them. This is called a prebond.

- Anneal the sandwich at 1100°C. At this extremely high temperature, all hydrogen diffuses off, and only silicon and oxygen remain. The oxide layers grow together and the two wafers are now bonded with strong Si-O-Si bonds.

When we require a membrane made from the device wafer, we have to etch and grind it down to the desired thickness. Then we polish the surface. This is called an SOI (Silicon on Insulator) substrate. SOI wafers can be bought readymade and treated like silicon wafers. The only difference is that there is an insulating oxide in between. They are used when a membrane with exact thickness and good crystal quality is needed, e.g. for high-quality pressure sensors. We will use SOI wafers for micromachined gyroscopes in Chapter 4.

2.4.2 Anodic Bonding

Fusion bonding is a high temperature process, and it is normally done at the beginning of a technology sequence, like in the case of making SOI substrates. What if we want to do full wafer bonding at the end of a process, e.g. to close a cavity with sensitive structures in it? Then we need a low-temperature wafer bonding method. You already know technology good enough to tell me what to do: when the energy does not come from heat, it has to come from an electric field. We cannot bond silicon to silicon this way, but silicon to glass. We use Pyrex glass, which has three important features:

- It contains a lot of sodium ions (Na^+). At room temperature, they cannot move and Pyrex is a good insulator. At elevated temperature >300°C, they become movable. Current transport by moving ions becomes possible.
- Its thermal expansion is very similar to silicon, and there is not much thermal stress.
- While silicon is hard and brittle, Pyrex is comparatively weak. It will deform a little before it breaks.

What do we do for anodic bonding? First, we do a Prebond by putting a clean, polished Pyrex surface on a clean, polished silicon surface. Then we put the sandwich on a hotplate as shown in Figure 2.16 and heat it to 450°C. The metallic hotplate acts as the lower electric electrode. We apply an electric current through the sandwich. Now the sodium ions move upward, away from the interface. The interface is depleted, and locally, a strong electric field is generated, due to the now missing charges of the sodium ions. This field presses the two wafers strongly together. When there is a small gap, then

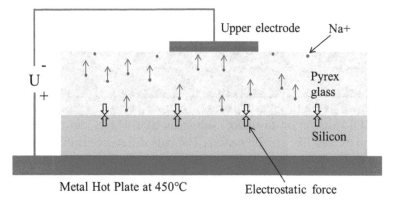

Figure 2.16 Anodic bonding of glass to silicon. At high temperature and electric field the sodium atoms in the Pyrex glass travel away from the interface. Strong local electrostatic attraction cares for tight bonding.

the weak Pyrex will bend a little and the gap vanishes. Due to this intimate contact, a strong bond is created.

Bonding of glass to silicon will be used for the pressure sensor. When we want to bond a silicon wafer to another one, how can we apply anodic bonding? The trick is to deposit a layer of Pyrex on silicon. We can bond this Pyrex and have a silicon–Pyrex-silicon sandwich. This can be used to hermetically seal the cavity of gyroscopes.

2.5 Making Structures

Making structures in a thin film or in silicon generally consist of two steps:

1. A photosensitive polymer layer (the resist) is deposited. The structures are written into this layer using optical lithography.
2. The structures are transferred into the thin film or the silicon by etching.

2.5.1 Lithography

Photolithography works like photography 20 years ago, when the films were made of celluloid. We use a photosensitive layer. The energy of a photon can break a chemical bond in the layer, which changes the molecule from insoluble to soluble. Then we remove the solvable molecules, which is called developing the picture.

The important process steps of lithography shown in Figure 2.17 are:

1. Making a <u>lithography mask</u>. A mask is a glass plate with a structured layer of chromium on it. The geometries are designed with a CAD tool. Then a mask shop structures the chromium layer using an electron beam writer for structuring.

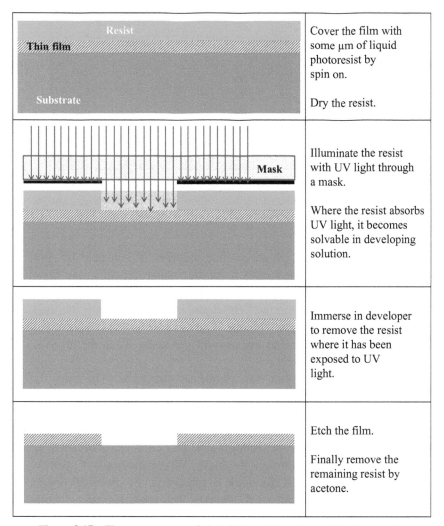

Resist Thin film Substrate	Cover the film with some μm of liquid photoresist by spin on. Dry the resist.
Mask	Illuminate the resist with UV light through a mask. Where the resist absorbs UV light, it becomes solvable in developing solution.
	Immerse in developer to remove the resist where it has been exposed to UV light.
	Etch the film. Finally remove the remaining resist by acetone.

Figure 2.17   The process steps of photolithography using positive photoresist.

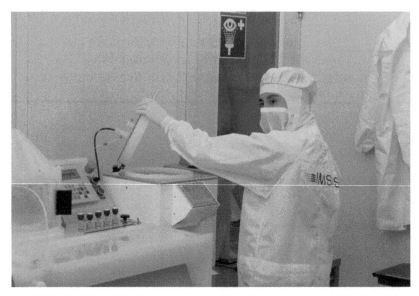

Figure 2.18 Working with the spin coater to deposit resist on a wafer. The cleanroom department for photolithography is illuminated with yellow light since white light contains short wavelength spectral components, which would expose the resist.

2. Making the <u>resist film</u>. A photosensitive polymer is deposited by spin coating. A wafer is placed onto a rotating stage (chuck), and then some drops of liquid resist are deposited in the middle (Figure 2.18). Afterward, the wafer is spun fast. The liquid is driven out by centrifugal force, and only a thin layer remains. The film thickness is controlled by the rotation rate, time, and viscosity of the resist. Then the resist is heated to remove the solvent.

3. <u>Align</u> the mask. Each mask has alignment structures. The mask is aligned on an x-y-Θ stage with respect to the previous layer using alignment microscopes.

4. <u>Expose</u>. For microsystems, the mask is usually placed directly on the wafer, which is called contact lithography. The photosensitive resist is illuminated by UV light through the mask. Where the mask is open (= the chromium thin film on the mask is removed), the chemical bonds are broken by the high energy UV photons. The resist is changed from insoluble to soluble.

5. <u>Develop</u>. With a developer, those parts of the resist which have been exposed are resolved. The underlying film lays open.

6. <u>Postbake</u>. Sometimes the resist is hardened by baking it at elevated temperature after being developed. This is done to make the resist more durable for the etching processes.
7. <u>Etch</u>. The underlying material is structured by etching. The parts that should not be etched are covered by the resist and thereby protected.
8. <u>Remove the resist</u>. The resist is dissolved by acetone. Alternatively, we can remove it in the gas phase in oxygen plasma (resist stripping).

This process is called the positive resist process, since the film will show a positive picture of the mask: where the mask is open, the film will be removed. Alternatively, there are also negative resists which work the other way round. They are initially soluble in the developer, but UV photons change them to insoluble. Then the film will be a negative picture of the mask. Normally we use the resist for etching, and then we remove it. But sometimes we need a film of polymer, and then we can leave the resist on the wafer and use it as a structural layer. An interesting layer is SU8, a negative resist based on epoxy that can be deposited and photo-structured in very thick layers up to 100 μm.

For microsystems, we generally use high-pressure mercury lamps at a wavelength of 365 nm. This could theoretically allow structures of $\leq$200 nm, but practically 1 μm is what we really achieve. When we make a line pattern, the distance between the two lines should be minimum, $\delta x = 5$ μm. When you want to be on the safe side, you should take a minimum of $\delta x = 10$ μm. Figure 2.19 shows a "lines and spaces" test structure. You can see that 2-μm spaces are good, but 1-μm spaces are not resolved by lithography in this case.

For microelectronics, this is obviously not good enough. CMOS technology normally uses excimer lasers like krypton fluoride at 248-nm wavelength or argon fluoride at 193 nm. Masks are not put in contact but lithography is done from the distance using a focusing lens. The feature size for many advanced CMOS processes is 50 nm, but smaller features with 16 nm have already been shown[6]. For nanostructures, E-beam lithography is used since electrons have a smaller wavelength than photons.

---

[6]16 nm-resolution lithography using ultra-small-gap bowtie apertures; Yang Chen *et al* 2017 *Nanotechnology* 28 055302

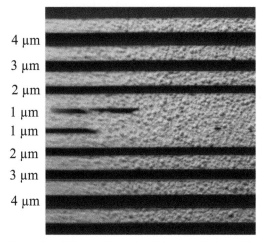

4 μm

3 μm

2 μm

1 μm

1 μm

2 μm

3 μm

4 μm

Figure 2.19   Checking photolithography and etching resolution with a lines and spaces pattern.
A 300-nm gold layer is structured by etching. We see that the 1-μm spaces are not resolved by lithography, the 2-μm spaces are resolved.
Figure by A. Schander.

2.5.2 Etching Fundamentals

Etching is the removal of the surface of a solid body by chemical reaction (dissolving) or by particles with high energy (sputtering). We define the etch rate R as the removed material thickness dy over time:

$$R = dy/dt \qquad (2.3)$$

We want to etch the film, but not the mask. The selectivity is the ratio of the rate of the desired etching of the film (R_1) and the undesired etching of the mask (R_2):

$$S = R_1/R_2. \qquad (2.4)$$

Wet chemical etchants usually etch in all directions with the same rate. This is called isotropic etching. Since the etching takes place isotropically, it will also remove material from under the mask, which is called underetching. If an etching process etches only in one direction, it is called anisotropic. The factor of anisotropy is given by the lateral etching rate R_l and the vertical etching rate R_v as:

$$A = 1 - \frac{R_l}{R_v}. \qquad (2.5)$$

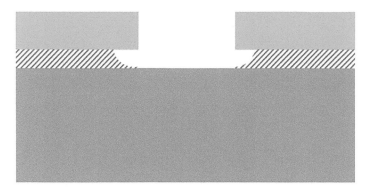

Figure 2.20 Isotropic etching: The etch rate is the same for every direction; the etch profile is round.

For wet chemical processes, we expect isotropic etching, with anisotropy being A = 0. This should cause a round profile, as shown in Figure 2.20. Underetching is 100% of the film thickness. Practically, it is even worse. To ensure that the film is completely removed on the whole wafer, we have to allow a little more etching time, doing some over etching. Vertically, overetching will not be a problem, if the selectivity versus the substrate is good. But laterally, underetching continues and we end up in an underetching >100% of the film thickness, as shown in Figure 2.21. We never will be really sure of the exact geometries of the structures. Hence, anisotropic etching processes as shown in Figure 2.22 are highly desirable.

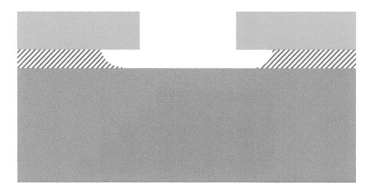

Figure 2.21 To be sure to remove all film material, we have to allow more etch time. This causes underetching.

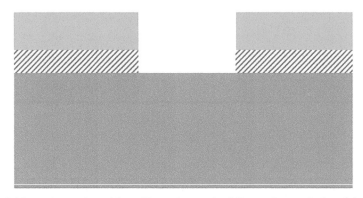

Figure 2.22 Anisotropic etching: The etch rate is different for vertical and horizontal direction. The etch profile is ideally a vertical line.

2.5.3 Anisotropic Wet Etching of Silicon

The first way to realize anisotropic etching uses the crystal structure of monocrystalline silicon. The method is very useful, but only in the monocrystalline silicon, not in thin films, also not in polycrystalline silicon. To understand the process, we have to know that in a monocrystalline material, the etching rate is different for different crystal orientation. For etching, we need to break the Si-Si bonds. This can be done using an alkaline solution from Potassium hydroxide (KOH). We first look at a 2-D model of a crystal in Figure 2.23. The crystal planes are described by the arrows normal to the plane. The notation is (hk) for a plane and [hk] for the vector normal to it. For the horizontal plane, the red arrow is [10] and I call it a (10) plane. A surface

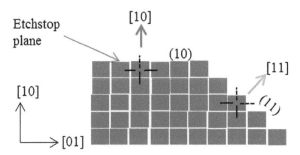

Figure 2.23 A two dimensional model for anisotropic etching of the crystal planes. To etch the (10) plane, we have to break 3 bonds simultaneously. This needs much energy and the etching rate is low. To etch the (11) plane, we only have to break 2 bonds simultaneously, and etching is much faster.

atom in this plane has 3 bonds to other atoms in the solid surface, which are called three backbonds, and it has one free bond, which is called a dangling bond. When I want to remove an atom by etching, then I must break 3 bonds simultaneously, and this plane is slowly etching. I call it an etchstop plane. The diagonal downward plane is described by the green [11] arrow, I call it a (11) plane. When I want to remove an atom here, I have only 2 bonds to break simultaneously. The (11) plane is more likely to be etched than the (10) plane, and the etch rate will be higher. This simple 2D model explains why different crystal planes show different etch rates.

In 3D, it is really difficult to envisage the planes. The best is to take a 3D crystal model for diamond in your hands (diamond and silicon have the same structure). Anyhow, let us give it a try and look what we can see from Figures 2.24–2.27:

1. In the volume, every silicon atom has four silicon neighbors arranged in a pyramid around it. This is because silicon is a tetravalent atom, like carbon.
2. A surface atom on the (111) plane has three back bonds and one dangling bond.
3. A surface atom of the (100) plane has two back bonds and two dangling bonds.
4. From this, we understand that (100) is easily etched, while (111) is the etch stop plane.

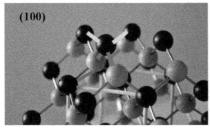 

When I put a plane of this unit cell on the table this corresponds to a (100) surface. The uppermost atoms have 2 backbonds each, shown in red. They have also two dangling bonds, shown in green. Accordingly, (100) is etching fast.

When I put a corner of the unit cell on the table, this corresponds to a (111) surface. Each surface atom has 3 backbonds (red) and one dangling bond (green). Accordingly, (111) is etching slow.

Figure 2.24 Photograph of a silicon crystal model. Each wooden sphere represents a silicon atom. The cubic unit cell is shown in yellow.

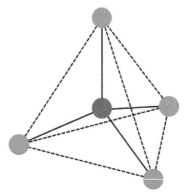

Figure 2.25   Silicon has four bonds. Every silicon atom is in the center of a tetraeder with a neighboring silicon at each corner.

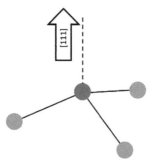

Figure 2.26   In the (111) plane, the Si atom in the center has three back bonds and one dangling bond. Hard to etch, this is the etch stop plane.

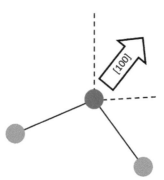

Figure 2.27   In the (100) plane, the Si atom in the center has two back bonds and 2 dangling bonds. Etches much faster than the (111) plane.

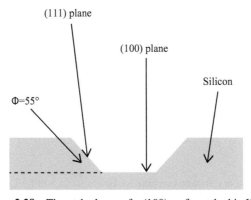

Figure 2.28 The etch planes of a (100) wafer etched in KOH.

5. In the cubic unit cell, [100] is a side of the cube, while [111] is a space diagonal. The angle Φ between them is:

$$\Phi = \arccos \frac{1}{\sqrt{3}} = 55°. \tag{2.6}$$

For KOH and silicon, (111) is a good etch stop plane; this implies there is a high selectivity between (111) and (100). The ratio is around 1:100, depending on temperature and concentration.

$$S_{KOH} = R_{(111)}/R_{(100)} = 1/100. \tag{2.7}$$

Figure 2.29 The etch planes of a (100) wafer etched in KOH, seen using an electron microscope.

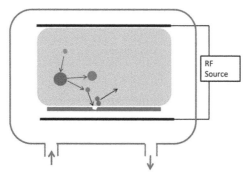

Figure 2.30   In principle, a dry etching machine looks similar to a PECVD reactor. With a vacuum pump and a gas supply, we realize a low pressure atmosphere of the etching gas. An RF source and electrodes are used to generate a plasma. The molecules of the etching gas are broken into reactive radicals and ions.

This selectivity can be used to etch straight planes into a silicon wafer. When the wafer has a (100) orientation (the wafer surface is a (100) plane), then the etch-stop planes will have an oblique angle of about 55°. When we want to make a silicon membrane for a pressure sensor, we will etch a (100) wafer with KOH. This will give the typical slopes as shown in Figures 2.28 and 2.29. These KOH slopes are typical for all membrane sensors until 2000, for most pressure sensors till today. For many people, Figure 2.28 is something like an emblem for silicon micromachining.

2.5.4 Dry Etching

An alternative to wet chemistry is etching in the gas phase using plasma. Figure 2.30 shows a reactive ion etching machine. It includes a gas supply, a vacuum system and an RF source. The source can do RF and DC bias. This allows us to switch from pure plasma burning to strong acceleration of ions and choose any mixture in between. Figure 2.31 shows the possibilities.

1. Chemical attack:
The plasma can break the bond of a gas molecule and produce reactive radicals. Using SF_6, a widespread etching gas, with argon as a carrier gas, the equation is:

$$e^- + SF_6 \rightarrow e^- + SF_5^* + F^* \tag{2.8}$$

The asterisk F^* indicates a reactive radical. These reactive radicals attack the surface chemically.

$$Si + 4F^* \rightarrow SiF_4 \tag{2.9}$$

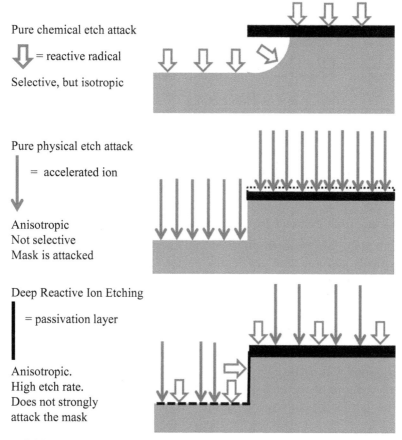

Pure chemical etch attack

⇩ = reactive radical

Selective, but isotropic

Pure physical etch attack

↓ = accelerated ion

Anisotropic
Not selective
Mask is attacked

Deep Reactive Ion Etching

▮ = passivation layer

Anisotropic.
High etch rate.
Does not strongly
attack the mask

Figure 2.31 Pure chemical etching by radicals, pure ion etching, and Deep Reactive Ion Etching (DRIE).

The reaction product SiF_4 is volatile and will be pumped away with argon. Chemical attack is highly selective but isotropic. This will result in a strong underetching of the mask, but the mask itself is not attacked since the chemical attack is very selective. The term plasma etching (PE) generally refers to a chemical etching process by radicals. It is used to structure films or to remove resist by stripping in oxygen plasma.

2. Physical attack: A particle can be ionized and accelerated. Using argon, the equation is:

$$e^- + Ar \rightarrow Ar^+ + 2e^- \tag{2.10}$$

When the argon ion hits the surface, it removes the material mechanically by sputtering off. In fact, argon ions are used as well as carbon ions. The process is not selective, but anisotropic since the mechanic momentum is a vector. Everything is sputtered off by brute force. Therefore, the mask is etched too and we cannot go very deep before the mask is destroyed.

3. Combined attack: Deep reactive ion etching

Our aim is to get anisotropic etching, but without the drawbacks of pure ion etching. This can be done using a combination of physical and chemical attacks[7]. The process is called RIE (Reactive Ion Etching) or DRIE (Deep Reactive Ion Etching). When you look at the RIE machine in Figure 2.30, it resembles a PECVD reactor: it has a gas system and an RF source. We could do PECVD deposition, and we just have to change the gas. What we do is to switch back and forth between the two processes:

1. Etching step: We use both physical and chemical attacks. We generate radicals and ions and accelerate the ions to hit the wafer surface vertically.
2. Passivation step: We deposit a thin passivation layer. We exchange SF_6 by C_4F_8. In the plasma, C_4F_8 is dissociated and a thin layer of CF_2 will deposit at the bottom of the etch groove and at its walls.

The trick is that the passivation layer of CF_2 is not etched by the radicals we use. The ions will destroy the passivation. After a passivation step, the wall and the bottom of the etch groove are passivated. Then we switch to etching. The ions hit the bottom, they destroy the passivation and the radicals can attack at the bottom to increase the depth of the etch groove. At the sides, there is a different story. The ions do not hit the side wall, and the radicals cannot destroy the passivation. Thus, the side wall stays passivated and will not be etched. What happens to the mask? The ions will attack it a little, but this is not a problem since the ion rate in DRIE is much smaller than in pure ion etching.

DRIE allows etching vertical walls in silicon. DRIE has been developed in the late 1990s. It is one of the most important technological innovations in silicon technology in the last 30 years. Aspect ratios of 6 to 10 are standard; much larger values can be achieved in research labs. Figures 2.32–2.34 give examples of DRIE-etched structures.

---

[7]Laermer, F. Schilp, A.: Method for anisotropically etching silicon. US Patent 5,501,893 (R. Bosch GmbH)(1994).

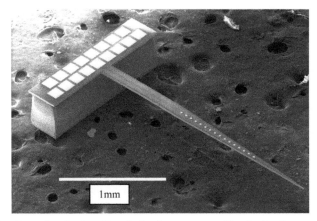

Figure 2.32 Example of a DRIE-etched microstructure. The device is a needle made of silicon with interconnects and contact electrodes on it. It is used as a neural implant to measure the activation potentials in the brain.
Chip by T. Hertzberg,
SEM by E. M. Meyer.

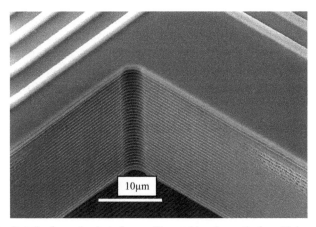

Figure 2.33 Details from the last figure. The etching is vertical; a high aspect ratio is realized. The ripple on the etched surface is generated by the switching between etching and deposition.

2.6 The Thermopile Process

The IR thermopile sensor is a good example to show how different technologies and processes previously explained are used to manufacture a device. In the following figures and in Table 2.2 we look at single steps and the complete sequence of the necessary technology process to fabricate the sensor.

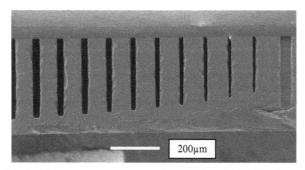

Figure 2.34   Small and large openings on a mask, etched by DRIE. The large openings show higher etch rate. This comes from a concentration of the electric field at larger open surfaces. This is called the loading effect, and we have to take it into account when designing a mask for DRIE.
Figure by C. Habben.

As a substrate, we use a silicon wafer which is double side polished. Doping and crystal orientation are not important. We want to make a membrane from silicon nitride. The wafer is new, and there is no temperature constraint and we can use LPCVD, since we know that it makes superior film quality. Unfortunately, LPCVD silicon nitride has extremely high intrinsic tensile stress and when deposited directly on silicon, the film might delaminate. Experience shows that nitride adheres much better to silicon oxide. What we do is to make a thin oxide first to improve adhesion, and to make the nitride next.

Step 1: Thermal oxidation of silicon: thickness 100 nm.

Step 2: LPCVD deposition of silicon nitride: thickness 300 nm. LPCVD is a double-sided deposition, and so the same two layers are deposited on the back of the wafer.

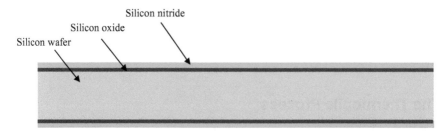

Step 3: is the deposition of polysilicon, also by LPCVD. Temperature will not be a problem since the two underlying films are deposited by high temperature. Adhesion of polysilicon on silicon nitride is also good.

Step 4: We have to structure the polysilicon to make the ingoing lines of the thermocouples. These lines are 10-μm wide, and so we cannot allow for much underetching. Hence, we chose dry etching (RIE) for structuring the polysilicon layer. This will provide precise definition of the sidewalls. But we pay a price for that: RIE etching of polysilicon is not selective to silicon nitride. There is no etch stop, and if we over-etch too much, then we will hurt the membrane. We have to control the etch rate extremely well, and so we can surely remove all the polysilicon but only slightly etch into silicon nitride.

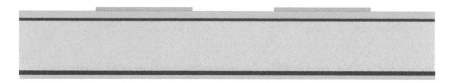

Step 5: is the deposition of aluminum, the material for the outgoing lines of the thermocouples and for the bond pads. We choose sputtering as method; it makes better films than evaporation. Substrate temperature during sputtering will be 250 °C, with no temperature issues. Adhesion of aluminum is good, on polysilicon as well as on silicon nitride. A possible danger is insufficient step coverage, but microscopy shows that step coverage is good (Figures 2.37–2.39).

Step 6: Aluminum is easy to etch wet chemically. Polysilicon and silicon nitride are both far more stable chemically than aluminum, and etch selectivity is good. Using a microscope, we check that underetching is small and borders are not too rough (Figures 2.38 and 2.39).

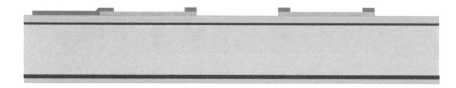

Step 7: is the deposition of PECVD silicon nitride. The thermopiles are structured by now. To protect them, we apply an insulating covering layer, the passivation. Unfortunately, we cannot use LPCVD deposition any more, as the aluminum would melt. We depend on using the PECVD layers. We know that this method, due to low temperature and low surface mobility, has only limited step coverage quality. The only way to guarantee a full passivation,

including steps, is applying a thick film. We choose a 2.5-µm layer of silicon nitride. Adhesion is good on all underlying films (silicon nitride, polysilicon, aluminum). PECVD has a small compressive film stress, which generally does not cause a problem since it is overcompensated by the high tensile stress of the membrane layer of LPCVD silicon nitride.

<u>Step 8:</u> is etching the passivation at the bond pads for electric contact. We use plasma etching, which has no selectivity versus aluminum. Again, we have to control the etch rate carefully not to hurt the metallization layer.

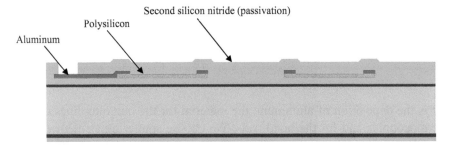

<u>Step 9:</u> Making the etch mask for removing the bulk silicon to set the membrane free. As the etch mask, we use the oxide nitride layers deposited in step 1 and 2. We turn the wafer upside down, and then we do photolithography. Etching of silicon nitride is done by plasma etching; the oxide layer is etched by wet chemical method using hydrofluoric acid (HF). Plasma etching of nitride will also attack oxide, but we do not care since the oxide has to go away anyway.

<u>Step 10:</u> removes the bulk silicon by DRIE. The etch time is ca. 90 min. We etch with vertical walls until etching stops at the silicon oxide.

<u>Step 11:</u> Finally, we apply the infrared radiation absorber. This is a film which is highly absorbing for all wavelengths from visible to 20 µm.[8] This can be

---

[8]M. Schossig, V. Norkus and G. Gerlach: Infrared Responsivity of Pyroelectric Detectors With Nanostructured NiCr Thin-Film Absorber. *IEEE Sensors Journal*, vol. 10, (2010).

done by evaporating a very porous layer from gold black.[9] This is a critical step since the film is very porous and mechanically weak. Structuring is done using a shadow mask.

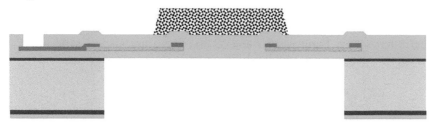

Table 2.2   Full process integration of the thermopile process. For a deposition step, we have to check for adhesion and for temperature problems. For an etching step, we have to check for selectivity and etch stop

No.	Step	Method	T [°C]	Adhesion	Etch Stop	Possible Problems
1	Oxide depo	Thermal oxidation	1100°	Ok	—	
2	Nitride depo	LPCVD	755°	Ok	—	Stress
3	Poly depo	LPCVD	620°	Ok	—	
4	Poly etch	RIE	—	—	No etch stop on nitride! Time control	
5	Al depo	Sputtering	250°	Nitride ok Poly ok	—	Step coverage
6	Al etch	PE	—	—	Nitride ok, Poly ok	Rough borders
7	Nitride depo	PECVD	280°	Nitride, Poly, Al ok	—	Pinholes, step coverage, stress
8	Nitride etch (Open bond pads)	PE	—		No etch stop on poly-Si! Time control	Etching Al
9	Nitride and oxide etch (Make KOH mask)	Plasma and HF	—	—	Not needed	
10	Si etch	DRIE	—	—	DRIE stops on oxide	Aspect ratio, inhomogeneity
11	Absorber	Evaporation	—	Can break off		Film is extremely porous

[9]Absorbing Layers for Thermal Infrared Detectors. W. Lang, K. Kühl, H. Sandmaier; Sensors and Actuators A, 34 (1992) 243–248.

This is an abbreviated process protocol, of course. To keep the list of our first full process short, I did not list the cleaning processes. I also left out the process measurement and quality control. When we discuss our second example, the silicon pressure sensor, this will be included.

After every process step, we have control success. One important method is optical microscopy, Figures 2.35 and 2.36. There are also electric test

Figure 2.35   Checking photoresist with the optical microscope.
At two places, the resist did not hold and broke off. Remove the resist, clean the wafer and start again.
Figure by A. Pranti.

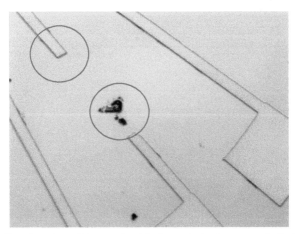

Figure 2.36   Checking the etching result with the optical microscope. Polysilicon lines of a thermopile after etching. There is dirt, and one interconnect is interrupted. This chip is lost. The dirt might be photoresist from a previous lithography not completely removed.

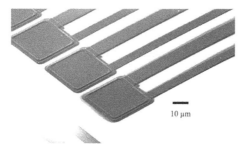

Figure 2.37 Details of a thermopile seen in a Scanning Electron Microscope (SEM). We see the contact pad of the polysilicon lines and the metal lines plus insulation layer. Figures by N. Hartgenbusch.

Figure 2.38 Upon observing closely, we see that the borders of the lines are not straight, but they show some roughness. Here this roughness is <<0.5 μm, this is ok. Roughness of etched lines must always be checked in technology.

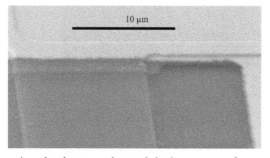

Figure 2.39 Observing closely, we understand the importance of step coverage. The last layer must cover the steps of the first polysilicon and of an insulation layer.
An experienced layouter will make the bondpads a little different in size for each layer, avoiding a double step this way.

structures, which we will look at when we discuss the pressure sensor. Figures 2.35 to 2.39 show process details made with optical and electron microscope. The explanations have been provided as figure legends.

2.7 The Recent Trend: Flexible Structures and Polymer MEMS

In recent times, flexible and stretchable micro sensors are a major focus of research. They are required whenever we want to integrate a sensor directly in a material, such as fiber compound material, plastics or elastomers. When I integrate a sensor in the structure material, I want to get data out of the material, but I do not want to downgrade the mechanical stability of the material. This is referred to as minimizing the 'foreign body effect' or the 'effect of the wound' in the material. The idea of small material integrated sensors is supported looking at sensing in nature. Figure 2.40 shows a pressure sensor from the skin of a mammal. In this case, it is taken from a rat, but human skin has similar sensors. This type of sensory receptor is called Merkel's disk, it senses mechanical pressure. Your fingertips have hundreds of Merkel's disks. The size of the receptor is around 50 μm, much smaller than today's silicon sensors. Yet, when we will discuss pressure sensors, we will see that the pressure sensor membrane can be as small as 60 μm, same size as a Merkel's disk. The real function can be provided in a very small size, but technology today needs additional space for chip, housing, and interconnects.

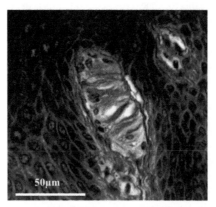

Figure 2.40   A Merkel's disk, a biological pressure sensor taken from the skin of a rat. Figure by Ursula Dicke.

The challenge is to do "function scale integration": throw away the ballast and integrate only what is really needed to perform the function.[10]

Can we realize such small sensors, which give us information about the interior of a material, but, due to small size and flexibility, do not disturb the matrix? We need highly flexible sensors, and they must have covalent bonding to the matrix material in order to prevent shearing off under load. Thin sensory foils can achieve this.

Such a sensor integration technology is desired for many sensorial problems. Here, I will only give a few typical examples.

The sensorial gasket:

Gaskets are low price components, but when they fail, the damage can be high. In critical applications, sensor equipped gaskets will improve security and allow predictive maintenance. Figure 2.41 shows a very small strain gauge made from a polymer-metal-polymer sandwich, embedded in a gasket. When the gasket is compressed, then the resistivity of the strain gauge changes. The elastomer of the gasket is a visco-elastic material. After a number of years, it may degrade, elasticity reduces, the material flows and

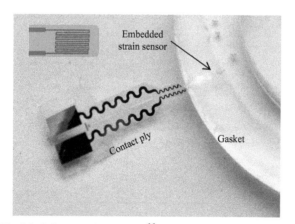

Figure 2.41 Flexible foil sensor in a gasket.[11] The sensor is a strain gauge that measures the deformation when the gasket is pressed. When the material of the gasket degrades, this deformation reduces. This indicates the danger of becoming leaky.

---

[10]Lang, W., F. Jakobs, E. Tolstosheeva, H. Sturm, A. Ibragimov, A. Kesel, D. Lehmhus, U. Dicke: From embedded sensors to sensorial materials-The road to function scale integration. Sensors and Actuators A: Physical, Vol. 171, Issue 1, November 2011, Pages 3–11 (2011).

[11]Embedded Strain Gauges for Condition Monitoring of Silicone Gaskets. T. Schotzko, W. Lang. Sensors, 14(7) (2014).

the gasket starts to leak. Simultaneously, the strain gauge will relax. This way the sensor can see the change in elastic behavior and warn us of leakage.

Sensors in fiber compound materials:
Fiber compounds are strong and light weight. They are a favored construction material for airplanes, wind energy rotors, and cars. Fibers of glass or carbon are embedded in liquid resin, which then polymerizes. Sensorial data are needed in production and in use:

- When there is mechanical shock such as a stone hitting the wing of an airplane, then the fibers may break inside the material. This cannot be seen from outside visually, but it can be detected using ultrasound transducers.
- Humidity can penetrate into the material, increasing weight and possibly reducing lifetime. We must measure the water content during operation.
- In production, the process of polymerization must be monitored: Is all the volume filled with resin or are there voids? Is the solidification complete? Polymerization changes the dielectric function of the material, and it is measured using embedded capacitive sensors. Figure 2.42 shows the process for a capacitive foil sensor. The foil is

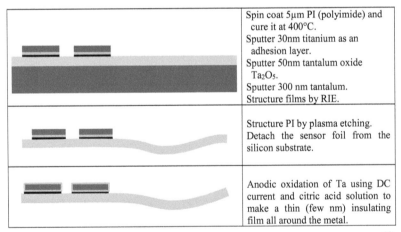

	Spin coat 5µm PI (polyimide) and cure it at 400°C. Sputter 30nm titanium as an adhesion layer. Sputter 50nm tantalum oxide Ta_2O_5. Sputter 300 nm tantalum. Structure films by RIE.
	Structure PI by plasma etching. Detach the sensor foil from the silicon substrate.
	Anodic oxidation of Ta using DC current and citric acid solution to make a thin (few nm) insulating film all around the metal.

Figure 2.42 Making of a flexible sensor as described by M. Kahali Moghaddam[12].

---

[12]Design, fabrication and embedding of microscale interdigital sensors for real-time cure monitoring during composite manufacturing. M. Kahali Moghaddam, A. Breede, A. Chaloupka, A. Bödecker, E.-M. Meyer, C. Brauner, W. Lang. Sensors and Actuators A 243 (2016)

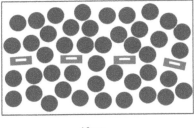

—— 10μm

Figure 2.43   The idea of sensor integration with minimal downgrading of the material.

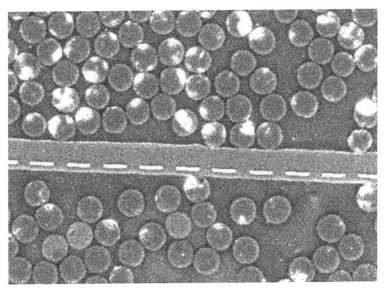

Figure 2.44   A foil sensor embedded in carbon fiber compound material. The diameter of the carbon fibers is 6 μm.

made of polyimide, the metallization is tantalum, and the insulation is tantalum oxide.

Figures 2.43 and 2.44 show a sensor embedded in carbon fiber compound material. This type of capacitive sensor is used for dielectric analysis of the material. Dielectric permittivity ε' and loss factor ε'' are measured as a function of frequency. During production we can measure the degree of polymerization: in liquid resin there are movable ions, which generate dielectric loss. When the material is fully polymerized, then the movement of ions is no longer possible. During operation, the sensor can detect penetrating humidity, since water has a much higher permittivity than polymer.

The challenge is to not only integrate the sensors, but also communicate with them. Integrating wires for interconnects is at least as challenging as integrating sensors. Transfer of energy by the electromagnetic field using the RFID technology does work[13], but antenna and electronics are larger than the sensors, and hence, the foreign body effect comes back again.

Material-integrated sensors and electronics[14] are a hot research topic now, since this is an important enabling technology for advanced digitalization and Industry 4.0.

Questions

- What methods do you use to deposit a gold thin film?
- Which method is better: sputtering or evaporation? Why?
- How do you deposit silicon nitride?
- Which one is better: PECVD or LPCVD? Why?
- Explain the process of photolithography.
- What resolution is possible with optical lithography?
- Which plane is the etch-stop plane in silicon? Why?
- How can I etch a vertical wall in silicon?
- How can I bond two silicon wafers?
- What does "anisotropy factor A = 1" mean?
- How do I measure the thickness of a film made of

  – Copper
  – Silicon oxide?

---

[13]Wireless Power Transmission for Structural Health Monitoring of Fiber-Reinforced-Composite Materials. M. Salas, O. Focke, A. S. Herrmann, W. Lang. *IEEE Sensors Journal*, Vol. 14, No. 7, July 2014.

[14]Material-Integrated Intelligent Systems: Technology and Applications. S. Bosse et al. ed., Wiley VCH (2018).

3

Force and Pressure Sensors

3.1 The Piezoresistive Effect

Imagine a wire of 1 m length, which is stretched by 1‰ to 1.001 m. The resistance of the wire will increase. First, 1‰ more length means 1‰ higher resistance. Then, due to stretching, there will be lateral contraction and the wire will become a little thinner, which will lead to an additional increase in resistance. Therefore, we should expect about 2‰ higher resistance. This is called the piezoresistive effect.

We define the piezoresistive coefficient k of a material as the relative change in resistivity per relative change in length

$$\frac{dR}{R} = k\frac{dl}{l} \tag{3.1}$$

To describe the deformation, we can also use the strain ε, which is the relative deformation:

$$\varepsilon = \frac{dl}{l}; \quad dR = Rk\varepsilon \tag{3.2}$$

Strain and stress should not be confused. Strain ε is purely geometrical and dimensionless. Stress σ is force per area and has the dimension of N/m^2. Strain values are generally small, and engineers often use the micro-epsilon: A strain of 1 $\mu\varepsilon$ means that $dl/l = 10^{-6}$.

The resistance R of a wire of length l, cross section A and volume $V = l \cdot A$ is given by

$$R = \frac{\rho l}{A} = \frac{\rho l^2}{V} \tag{3.3}$$

95

When we stretch the wire, the change in resistance becomes

$$dR = \frac{\partial R}{\partial l}dl + \frac{\partial R}{\partial V}dV + \frac{\partial R}{\partial \rho}d\rho = 2\frac{\rho l}{V}dl - \frac{\rho l^2}{V^2}dV + \frac{l^2}{V}d\rho \qquad (3.4)$$

$$\frac{dR}{R} = 2\frac{dl}{l} - \frac{dV}{V} + \frac{d\rho}{\rho} \qquad (3.5)$$

The first two terms describe the change of resistance due to the change of geometrical shape. The last term is the intrinsic effect, i.e. the change of the material parameter ρ induced by the strain.

Metals: For a metal, we can assume that a small deformation does not change the specific resistivity ρ and thus the third term vanishes. Next, we assume that the volume does not change during deformation, i.e. the decrease in cross section cancels the increase in length. Then, the second term also vanishes and we find $k = 2$. Yet we know from technical mechanics that the volume of a deformed body will change a little bit, and there will be a contribution of the second term. Practically, we assume $k = 2$ for all metals as a first guess, but we have to bear in mind that this really is a rough approximation.

Semiconductors: In semiconductors, mechanic deformation may change the level of the energy bands. Then, electrons are shifted from the valence band to the conduction band and vice versa, the specific resistivity ρ changes and the third term in Equation 3.5 becomes predominant. For silicon, we find k values between -100 and $+100$. The sign depends on doping and on the configuration. As for thermoelectricity, the band structure provides a much higher sensor effect in semiconductors than in metals. When we discuss silicon pressure sensors, we will come back to the piezoresistive effect specifically in silicon.

Composite materials: There are a number of composite materials that are essentially conductive particles embedded in a nonconductive matrix:

- Printed films (screen printing, inkjet printing or aerosol printing).
- Polymers with added conductive fillers such as carbon nanotubes or graphite.

Figure 3.1 shows a network of conductive filaments in an insulating matrix. Current transport through this material is possible, but strongly inhibited. In the figure, there is a current path where the conductive filaments touch,

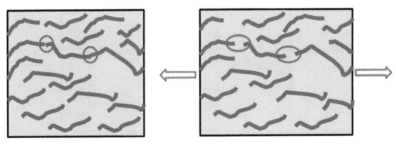

Figure 3.1 Conductive filaments in an insulating matrix. There are current paths where the filaments touch (red circles). When the material is strained, the filaments lose contact and resistivity increases significantly.

or almost touch. When they approach closer than 2 nm, current transport is possible due to tunneling of electrons. Therefore, we expect that the material will not be insulating, but specific resistivity will be high. Now imagine the same material strained, as in the right-hand side of Figure 3.1. The ends of the filaments are torn apart, the conductive path is almost or totally interrupted and resistivity will increase significantly. Therefore, we can expect a high piezoresistive coefficient. For graphite–polymer composites,[1] it can be as high as $k = 100$.

Furthermore, we may expect a highly nonlinear piezoresistive behavior. For low deformations, there is only a small effect. When strain increases and the conductive particles start to lose contact, resistivity increases significantly. This is called the percolation threshold.

A classic use of this principle is the carbon microphone, as shown in Figure 3.2. It used to be in every telephone receiver when I was a child in the 1960s. It is a resistor made of loose charcoal granules between two electrodes. Sound pressure will press the granules together, thereby generating more contact points and improving conductivity. This is history, but the same effect of touching particles is found in screen-printed metals, which show piezoresistive coefficients up to 15. These materials are of high technological interest, since they can be deposited by printing. Printed materials for sensing are subject of intensive topical research. An experimental device of a fully printed sensor is given at the end of this chapter.

---

[1]Piezoresistivity, Strain, and Damage Self-Sensing of Polymer Composites Filled with Carbon Nanostructures. F. Aviles, A. I. Oliva-Aviles, M. Cen; Advanced Engineering Materials (2018).

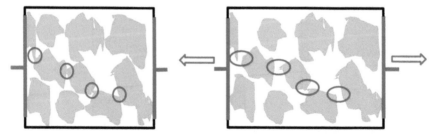

Figure 3.2 Piezoresistivity of conducting particles embedded in an insulating membrane. Examples are charcoal granules, screen-printed metals. Due to the percolation threshold, the piezoresistive effect can be very strong, e.g. Ni clusters in sputtered carbon show $k = 9$.[2]

3.2 Strain Gauges and Force Sensing

To measure force, we first need a mechanical structure that responds to the force with a known deformation. This may be a leaf spring or cantilever, as shown in Figure 3.3. Then, we apply a resistor on this probe body, and mostly we will use a meander made using thin-film technology. This is called a strain gauge, and in the force sensor community, the piezoresistive coefficient k is also called the gauge factor. When the tip of the cantilever is pressed down, the strain gauge on the cantilever undergoes tension and the resistivity increases. When temperature increases, the resistivity also increases, and to separate the two effects, we use a Wheatstone bridge. We could apply two resistors R_1 and R_4 on top of the cantilever, and two others R_2 and R_3 on the rear side. While R_1 and R_4 undergo tension and increase in resistivity, R_2 and R_3 undergo compression and decrease in resistivity. This causes an output in the bridge voltage. When temperature increases, all four resistors

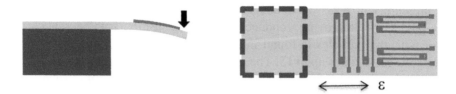

Figure 3.3 A cantilever bend down with a metal strain gauge.
In the figure, we see four strain gauges: two parallel to the strain, which will change in resistance, and two perpendicular to the strain, which will not change.

---

[2]https://www.celago-sensors.de/

will increase resistivity, which will not affect the bridge voltage. In this way, the temperature cross-sensitivity is cancelled.

This is working, but cumbersome, since we need resistors on both surfaces of the cantilever. There is simpler way, when we apply four resistors with different orientations. Two resistors are parallel to the strain, and they will be affected by strain. Two other are perpendicular, and they will not be affected. It is simpler to realize, but sensitivity will be smaller, since only two resistors are changed by the deformation.

The advantage of a Wheatstone bridge is the compensation of the temperature effect. A temperature increase will lead to an increase of all four resistors. Thus, the bridge voltage will not be affected.

How do we apply the resistors? There are two methods:

1. Thin films on a flexible foil, which is glued onto the part whose surface pressure is to be measured.[3] Examples are gauges glued onto a weighting spring in digital bathroom scales or gauges glued on the chassis of a car for testing the driving dynamics. The disadvantage of this technique is that thickness of the bonding layer varies, which can compromise the reproducibility of the measurement.
2. Thin film directly deposited on the material. An example is a pressure sensor with thin-film gauges on a steel membrane.

An advantage of strain gauges is the high resolution, which can reach 1×10^6. Further advantages are the easy application by gluing the sensors onto something and the affordable price.

Disadvantages are the large area of the meander and the small coefficient of $k = 2$. Furthermore, the baseline stability of the Wheatstone bridge circuit is limited. You have to recalibrate zero each time, when you start the device. In the digital scale, this is done when you push the On button: the system is started and it is automatically calibrated to zero.

Figure 3.4 shows a strain gauge on a steel spring which is taken out of a kitchen scale.

3.3 Pressure Sensors: Dimensions, Ranges and Applications

What is pressure? Pressure is the force a fluid imposes on a surface. The dimension is the Pascal: $1 \text{ Pa} = 1 \text{ N/m}^2$. Engineers often use the dimension

---

[3]https://www.hbm.com/en/0014/strain-gauges/

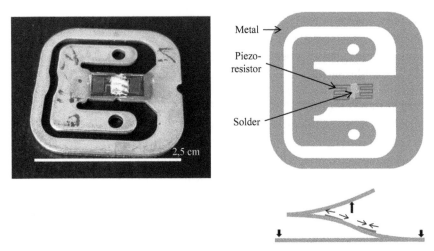

Figure 3.4   A piezoresistive sensor from a kitchen scale. The scale has one of these at each corner. On a metal plate, there are two metal meanders as piezoresistors: one is compressed, and the other is elongated. In the right-hand side, the deformation mode of the plate is shown.

bar: 1 bar = 10^5 Pa, very roughly the pressure of the atmosphere at sea level. Correspondingly 1 mbar equals 100 Pa or 1hPa (hectopascal).

What do we need pressure sensors for? And what ranges do we need? The following list is only a small excerpt of possible applications.

- A range of 1 bar sensors are used to measure altitude or atmospheric pressure. High-end smartphones have a high-precision pressure sensor for atmospheric pressure. When altitude is known from GPS, the atmospheric pressure is used for weather forecast. The other way round, altitude information can help GPS. GPS today can tell you in which building you are, but it cannot tell you the number of the floor. This would be important to help you find your way and especially in rescue situations. A high resolution pressure sensor can see that you went up 6 m after entering the house, thus, its third floor. Although, cars need to know atmospheric pressure to optimize the motor management.

- Hydraulic and pneumatic systems must be controlled. Pressure sensors can verify the operation of a pneumatic system, and they can warn if there are leaks. In large industrial plants, leakage in pneumatic piping is a major waste of electric energy. Pneumatics is the driving of devices using compressed air, typically the line pressure is up to 6 bar. Hydraulics uses fluid power of water or oil, and applied pressure is higher, in the range of 60 bar. Air-conditioning systems also use compressed fluids, typically at 20 bar.

- A major use in cars is the measurement of tire pressure using 5 bar sensors. When a car tire does not have enough pressure, the friction between the tire and the road increases. Experts estimate an increase of fuel consumption by 1% if the pressure is 0.2 bar too low.[4] Furthermore, the continuous kneading ruins the rubber, the lifetime of the tire is reduced and even accidents may happen. In Germany, continuous monitoring of each tire has been obligatory for new cars since 2014.
- Medicine technology wants to measure several pressures in the human body. The classical way to measure blood pressure is to use an inflatable cuff, which is inflated until the blood stops flowing. Many automated systems are around, most of them using piezoresistive silicon sensors. Integration of very small pressure sensors on syringes has been shown, but did not establish in medicare yet.

  The eye is filled with a liquid called aqueous humor. It is continuously produced and drained, at a pressure of about 1300–2700 Pa. If your body fails to control this, then you have a serious illness called glaucoma. The pressure increases, which may damage the optical nerve and can even cause loss of vision. The intraocular pressure is measured mechanically (tonometry), but ophthalmologists would like to have an implantable system for continuous monitoring. Systems have been developed, and in 2014, first implantation has been made.[5] The major challenge is miniaturization (lack of much space in the eye) and power supply. The implant may be powered from the electric field of a reading device (RFID principle), but this means that the system must have a sensor, communication electronics and antenna integrated.

  A third advancing medical application is brain pressure. Also, the liquor in the brain is continuously generated and drained. If it increases pathologically above its level at 1300 Pa, the blood cannot feed the brain any more with oxygen. RFID-powered sensors have been developed and applied also here.[6]

  All pressures in the body are measured as differential pressure versus atmospheric pressure. This is a problem for the integration of pressure sensors in the body. We cannot bring the atmospheric pressure to the

---

[4]https://www.adac.de/infotestrat/tanken-kraftstoffe-und-antrieb/spritsparen/sparen-beim-fahren-antwort-8.aspx

[5]http://implandata.com/erster-patient-in-implandatas-argos-02-studie-fuer-implantierbaren-augeninnendrucksensor-eingeschlossen/

[6]https://www.fraunhofer.de/de/presse/presseinformationen/2014/Januar/wasserkopf-sensor.html

other side of the membrane, so we have to measure two absolute pressures, one in the body and the other outside, and then subtract. High precision, good long-term stability and good linearity of our sensors is a prerequisite for this.

- Let us have a look at the other end of the scale: high pressure. For process control, ranges go up to some 100 bar, and in extreme cases, to 500 bar. This can be done by piezoresistive sensors, with metal membranes or with silicon membranes. Modeling of the sensors becomes challenging, because at 500 bar, not only the membrane is bend, but everything, including the housing of the sensor. The highest range I had to deal with was 1200 bar for deep sea level.

- Finally, let us look at the other end of the scale: extremely low pressures. The first application is vacuum technology, a rough vacuum will have about 1 Pa pressure, and a good vacuum system should go down to 10^{-4} Pa and lower. Vacuum sensors measure the thermal conductivity of air, which must break down if the air is pumped out. For very low pressure, they measure the ionization of air. This is done in large glass bulb devices. Micro sensors with membranes cannot do this task.

- The second low-pressure application is sound. In fact, sound propagates at atmospheric pressure, but the oscillating pressure of sound waves is very faint. The auditory threshold is about 20 μPa, and more than 60 Pa will cause pain and serious injury. Discussion about vacuum sensors and microphones needs a lot of specific background, and we will not discuss those devices in this book.

3.4 Piezoresistive Pressure Sensors

3.4.1 Steel Membrane Sensors

The transduction structure of most pressure sensors is a membrane that is deflected by the pressure. It transduces the pressure in the fluid medium into a mechanical strain, which can be measured by the piezoresistive effect. There are essentially two types: steel membranes with metal thin-film resistors and silicon membranes with silicon resistors defined by local doping.

Let us first consider steel membrane sensors. Figure 3.5 shows a membrane deflected by a pressure difference.

A pressure sensor membrane is approximately 50 μm thick. When deformed, it resists the deformation by its stiffness. Deformation will be

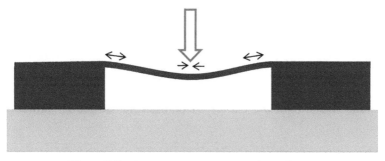

Figure 3.5 A pressure sensor on a steel membrane.

concave in the middle, but convex at the edges. In this way, on the upper surface, we find compressive strain in the middle and tensile strain at the edges. It is important to look closely at the edges! There is no sharp bend, but the membrane is smoothly bent down.

What we need is to place four resistors for a Wheatstone bride, two of them increasing and two decreasing in resistivity. Regarding Figure 3.5, we know what to do: put two resistors in the middle, the other two close to the edge as shown in Figure 3.6.

Understanding the layout, the technology of steel membrane pressure sensors is straightforward:

- Prepare a steel membrane on a circular rim, like a small can.
  This can be done by deep drawing of steel using a punch.
- The membrane is electrically insulated by PECVD silicon nitride.
- Metal is sputtered and structured by lithography and etching.

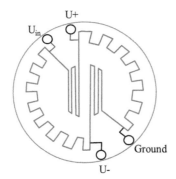

Figure 3.6 Pressure sensor, seen from above.

- A second silicon nitride is applied for passivation.
- Finally, a second lithography and etching open the bond pads.

The technology is well known, but unfortunately, lithography has to be done for each device separately. This makes the devices expensive. Yet, due to the superior robustness of the steel membrane, for many high-end applications, steel membrane sensors are the choice. A commercial device is shown in Figure 3.7.

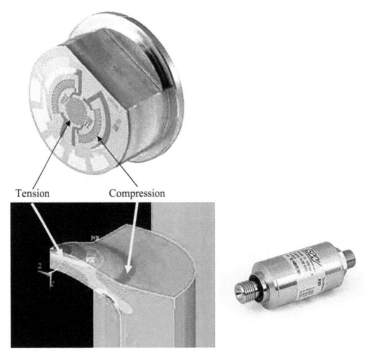

Figure 3.7   Piezoresistive pressure sensor on a steel membrane. The diameter of the membrane is 3 mm. This sensor is used in hydraulic systems.
Left: The sensor element
Bottom right: Housed sensor
Bottom left: A finite element simulation shows the different strain in the center and at the periphery of the membrane.
Pictures with kind permission of Wolfgang Wiedemann, Sensortechnik Wiedemann GmbH.[7]

3.4.2 Silicon Membrane Sensors

In fact, we would like to fabricate many devices on a single wafer, and thus we proceed to the second type, the silicon pressure sensor. First, we have to consider the piezoresistive effect in silicon. When a semiconductor is deformed, the energy levels in the band structure change. In this way, the specific resistivity ρ changes, and the intrinsic term in piezoresistivity becomes predominant. Depending on the doping and the geometrical configuration, the coefficient ranges from $k = -100$ to $k = +100$. What does "geometrical configuration" mean? There are two possible geometrical ways to align a piezoresistor with respect to the strain:

1. The current can be parallel to the strain. This is called a longitudinal configuration. Up to now, we have always been assuming longitudinal configuration for strain gauges.
2. The current can be perpendicular to the strain. This is called a transversal configuration. In silicon, the transversal configuration shows a strong negative piezoresistive effect, which means that resistivity is diminishing with strain.

When we experimentally measure the piezoresistive coefficient k, for p-doped monocrystalline silicon, we find:

Longitudinal: k∼ +120
Transversal: k∼ −110

This awakens our interest because we are always looking for opposing effects, in order to use Wheatstone bridge circuits. Therefore, we put two longitudinal resistors and two transversal resistors on one membrane, as shown in Figure 3.8 and we connect them in a bridge circuit. This is the piezoresistive

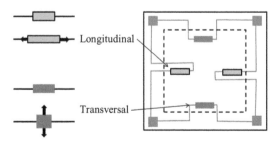

Figure 3.8   The piezoresistive pressure sensor on a silicon membrane. Two longitudinal (k positive) and two transversal (k negative) resistors form a Wheatstone bridge.
Note that the mechanical strain always points outside in, for all four resistors.

pressure sensor with a silicon membrane, maybe the most famous layout idea of the MEMS technology.

For polycrystalline silicon, the *k* values are much smaller. Therefore, it seems to be the best choice to make a pressure sensor with piezoresistors out of monocrystalline silicon. In this way, we can realize a bridge circuit with a clever combination of transversal and longitudinal resistors, and we can make use of the high piezoresistive effect in monocrystalline silicon, provided by the band structure deformation induced by the strain. But we pay a price: the band structure also changes with increasing temperature. The resistivity and the piezoresistive constant *k* will be functions of temperature. For the silicon pressure sensor, we will need temperature compensation, and when we want an accuracy of 1% or better, we will also need single sensor calibration.

3.5 From the Wafer to the System – Technology of the Silicon Pressure Sensor

3.5.1 Silicon Technology

Figure 3.8 shows the top view of a piezoresistive sensor in silicon technology. This type of sensor was developed around 1970; it has been the first silicon sensor produced in a large scale and it still is the paradigm of a micro sensor and one of the commercially most important micro devices. For this reason, we want to analyze its making in detail.

We need a membrane made of monocrystalline silicon. The piezoresistors are made in this membrane, but they have to be electrically insulated from each other. This can be done by local doping. When the membrane is n-doped and the piezoresistors are p-doped, there is no current flow from one resistor to another, since there are two opposed diodes in between, of which one always will be in reverse. How wide and thick must the membrane be? A typical value of a silicon pressure sensor is a width of 1 mm and a thickness of 30 μm.

The piezoresistors are defined by local p-doping of the silicon by ion implantation in an n-silicon membrane. The metallization is made by thin-film aluminum. The membrane is structured by KOH etching of the bulk silicon. When the silicon chip is finished, it is bonded on a chip of Pyrex glass and then mounted in a metal or plastic housing. The Pyrex glass is weaker than silicon and reduces the thermal stress between the silicon chip and the metal of the housing.

We start with an n-doped wafer, since the piezoresistors will be p-type, and we need a diode between the resistor and the membrane, otherwise the resistors would be short-circuited. The orientation of the wafer is (100) to allow KOH etching of the membrane.

Step 1: Mask 1 has only alignment crosses, which are etched into the silicon. This is needed because step 2 is doping, but doping is not visible, and in this way, step 2 will not make any structure we could use to align the following layers.

Step 2: Then, we can do local ion implantation. We spin on photoresist and perform lithography. Mask 2 has openings for the four resistors. We implant boron ions, and in the resistors, the doping changes from n-type to p-type. Where the resist is not opened, the ions stick in the resist and are removed with it.

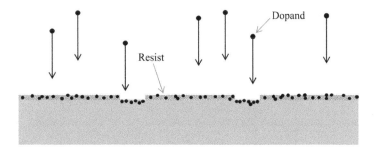

We must make sure that the p–n junction between the resistors and the membrane make good diodes. Between two resistors, there are always two diodes, one of which is in reverse. If the diodes do not work, creeping currents between the resistors will destroy the sensor effect. This can happen if the implantation dose is badly controlled or if there is ionic contamination in the clean room.

Step 3: Now, we need an insulating layer before we can make metallic interconnects. From the thermopile process, we know that a sandwich of silicon oxide and silicon nitride will be a good choice. High temperature is no problem at this stage, and we can use LPCVD for both layers, which result in high film quality. A further advantage is that LPCVD is a two-side deposition, and we can use the insulation layers as a KOH etch mask later. Why do we not use thermal oxidation here? This would be a mistake, since we would oxidize the piezoresistors we have just been implanting. (In the figures, we use only one green layer for both LPCVD films).

<u>Step 4</u>: Lithography with mask 3 provides two contact holes for each resistor.

<u>Step 5</u>: We etch nitride by RIE. There is no etch stop versus oxide, but this is not a problem, the oxide has to go away anyway.

<u>Step 6</u>: We etch oxide wet by HF. This provides a good selectivity versus with silicon.

<u>Step 7</u>: The interconnects for the Wheatstone bridges are made from sputtered aluminum. Surface adhesion is good. The silicon has to be cleaned from any oxide before sputtering; otherwise the contact resistance may increase.

<u>Step 8</u>: We structure the aluminum by wet etching (mask 4). We have to check line width for underetching. We also have to check contact resistance and the voltage/current curve of the contact. If it is linear, then we have an ohmic contact and everything is ok. If it is not linear, then we may have a Schottky contact, which causes deviation of the sensor characteristics. For these measurements, we use a wafer prober and the Kelvin test structure explained below.

<u>Step 9</u>: For passivation, we use PECVD silicon nitride. LPCVD is no longer feasible, unfortunately, due to the aluminum, which would melt.

<u>Step 10</u>: The front side is finished and we can make the membrane. We do lithography on the back of the wafer with mask 5 and then etch nitride and oxide as in steps 5 and 6.

<u>Step 11</u>: Next, we etch the wafer in KOH. The wafer surface is a (100) plane and will be etched. The etch stop planes (111) are in an oblique angle. The wafer is 525 μm thick, and we want a 30 μm membrane. The etch rate of the KOH is not constant, so we will not get exactly 30 μm in one etch step. We

make two steps with a measurement in between. We calculate the time for 475 µm, which leaves about 50 µm. The etch rate is 1.4 µm/min, so it is a very long etching time of more than 5 h. We take the wafer out, rinse it and measure the membrane thickness with a mechanical caliper. Then, we go back into KOH and etch the rest. Finally, we have to remove the nitride/oxide mask.

There are more refined methods to control thickness without intermediate measurement, such as using an SOI wafer or electrochemical etch stop at a p–n junction. These methods allow better control of the membrane thickness, but on the other hand, they are more difficult and expensive.

<u>Step 12</u>: We need electric access. With mask 6 and plasma etching, we open the bond pads. There is no selectivity versus the aluminum, so we have to control etch time carefully.

<u>Step 13</u>: Finally, we anodically bond a wafer of Pyrex glass on the rear side. This is to relieve thermal stress from the housing. If we would mount the silicon directly on a metal housing, then there would be thermal mismatch. The sensor cannot differentiate this from a pressure signal. The weaker glass relieves the stress. When making a differential pressure sensor, there is a hole in the Pyrex made by ultrasound grinding.

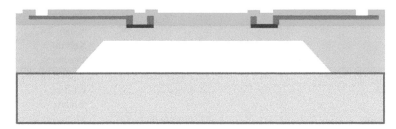

Table 3.1 summarizes the arguments for full process integration. For each step we have to check for compatibility with previous layers. For deposition we look at adhesion, for etching we look at selectivity.

3.5.2 Electric Test Structures

After every process step, we have to control success. To do this, we make some test structures on our mask. For measuring electric test structures, we use a wafer prober. The wafer is put on a stage, which can be moved in x, y and z directions. We have needles on micromanipulators, which we use to contact the bond pads of the test structures. We measure the test structure, then we lower the wafer a little to break contact, we move the wafer x and

Table 3.1 Process of the piezoresistive silicon pressure sensor

	Process Step	Selectivity	Adhesion	Possible Problems	Test Methods
1	Lithography				
2	Implantation			p–n junction not insulating	Electric
3	Oxide and nitride LPCVD		Oxide ok Nitride ok	Adhesion nitride on Si poor, use oxide between	Interferometer for film thickness
4	Lithography			Misalignment, resolution	Microscope, lines and spaces
5	Etch nitride PE	Low, but not critical			
6	Etch oxide HF	good			
7	Al deposition, Sputtering		Ok	Contact resistance, homogeneity	Resistivity: Four top measuring
8	Al etch wet	good		Underetching	RS: Greek Cross Underetching: 4-point structure Via resistivity: Kelvin structure
9	Nitride PECVD		Ok	Pinholes, step coverage	Interferometer for film thickness
10	Structuring the mask for KOH etching			Misalignment	Microscope
11	KOH etch			Etch rate not constant	Mechanical caliper
12	Open nitride	Poor, time etch			
13	Anodic bonding			Unbonded areas due to dust particles	Optical inspection
14	Saw			Braking membranes	

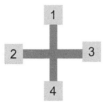

Figure 3.9   Greek cross for sheet resistance. Current in 1→2; Measured voltage: 3–4.

y to the next test structure and lift it to make contact and measure again. An automated wafer prober can do hundreds of measurements in a few minutes.

The <u>Greek cross</u> in Figure 3.9 measures the sheet resistance $R_\square$ ("R-square"), which is the specific resistivity ρ over the thickness D:

$$R_\square = \frac{\rho}{D} \tag{3.6}$$

Sheet resistance is the resistance one square of our film would have, independent of its size. How do we measure $R_\square$ from the creek cross? We drive a measurement current from port 1 to port 2, and we measure voltage between ports 3 and 4. The resistivity will not correspond to one square since we drive the current around the corner, but there is a strict relationship, which can be calculated:

$$U_{34} = 0.22\, R_\square\, I_{12} \tag{3.7}$$

The advantage is that this measurement is not dependent on the specific line width, so underetching will not disturb the measurement. If we want to be precise, then we do several measurements by cyclic permutation of the ports.

Sheet resistance is also very useful to calculate the resistivity of a line when we look at a mask. The line has a length of L and a width of W; then, we consider it as having $n = L/W$ squares. The resistivity is given by

$$R = \frac{\rho}{D}\frac{L}{W} = R_\square n \tag{3.8}$$

This allows us to measure the line width using a combination of Greek cross and four-point structure, as shown in Figure 3.10. First, we measure the sheet resistance using ports 1 to 4. Then, we drive a current from 1 to 6 and we measure the voltage drop between 4 and 5. Knowing the distance L from 4 to 5 and $R_\square$, we obtain the width W of the films.

$$W = \frac{\rho}{D}\frac{L}{R} = \frac{R_\square}{R}L \tag{3.9}$$

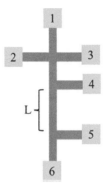

Figure 3.10   Greek cross combined with four-point structure for sheet resistance and line width. Current in 1→6; Measured voltage: 4–5.

Comparing the measured width W with the mask, we obtain the underetching, an important parameter for etching of metals.

Next crucial parameter is the resistance of the contact between the metal and the doped silicon of the piezoresistors, measured using a Kelvin structure, as shown in Figure 3.11. We drive a current from I_{in} to ground and measure the voltage drop between U_+ and U_-. Line resistances in the current path will generate a parasitic voltage drop, of course, but due to the four-lead configuration shown in Figure 3.12, this does not affect the measurement. We measure voltage drop as a function of current, and we also inverse the current. We should find a straight line with small slope. When the slope increases, we have increased contact resistance, maybe there is a little oxide between the silicon and the metal and we should improve cleaning before sputtering. When the curve becomes asymmetric with respect to the inversion of measurement current, we do not have an ohmic contact, but a Schottky

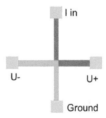

Figure 3.11   Kelvin structure for contact or via resistance.
Red: bottom layer
Green: top layer
Ochre: bond pads

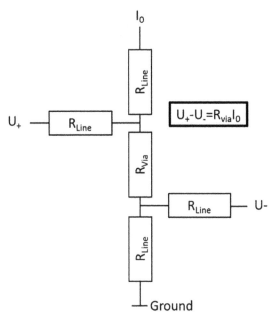

Figure 3.12 Four-lead measurement in the Kelvin structure: The voltage between U_+ and U_- is only given by the current and the resistance of the via, R_{via}. The line and contact resistances R_{line} are not influencing the measurement.

transition – or something between. Then, we should increase doping slightly or introduce a local contact implantation to our process.

3.5.3 Mounting and Housing Technology

With anodic bonding, the silicon technology is finished and we take the wafer out of the clean room. We saw it into single chips, and all the following processes of mounting and housing are single-chip processes. This is often referred to as "cycle 2" of sensor technology. Figure 3.13 shows a typical configuration of a hybrid integrated chip. The substrate is a printed circuit board. It has interconnects made of thick metal film, typically made by screen printing. The process of mounting the chip on the board is called die bond. A dispenser will place a little adhesive on the board, and then the chip is taken by a pick-and-place machine using vacuum tweezers and placed on the board. An ASIC for the sensor electronics is put at the side.

Electric contact is made by thin wires of gold or aluminum, as shown in Figure 3.14. This process is called wire bonding. The connection of the wire to the bond pad is done by ultrasonic welding:

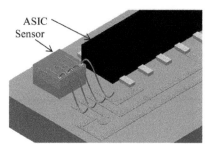

Figure 3.13   In hybrid integration, a sensor and an electronic chip are mounted side by side on a board and connected by bond wires.
Figure by G. Dumstorff.

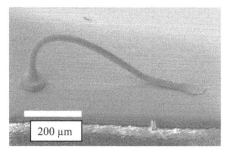

Figure 3.14   A bond loop made from a thin gold wire. Left: the ball bond; right: the wedge bond.
SEM by E.-M. Meyer.

- A thin gold wire (~30 µm diameter) is guided by a capillary tube.
- We use a short electric spark to melt the end of the wire locally, and the molten gold will contract and form a small ball.
- This ball is pressed down by the capillary tube on the bond pad. We vibrate the tube with a piezoelectric ultrasound actuator. The ball and the bond pad rub against each other and, very locally, a high temperature is generated. The metals fuse by ultrasound welding.
- We withdraw the tube, more wire is drawn out of the tube, and we move the tube to form the elegant loop, as shown in the figure.
- We press the wire against the interconnect on the board and activate the ultrasound again. Then, the wire is locally welded, in the form of a wedge. Finally, the wire is torn off.

When you look at an automated Ball-Wedge bonder, the process reminds you of a sewing machine.

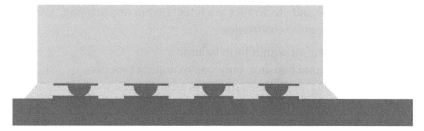

Figure 3.15   Flip chip bonding of a chip to a board.

Note that there are three processes called bonding, and they are very different:

- <u>Wafer</u> bonding is a full wafer process bonding silicon to silicon (fusion bond) or to Pyrex glass (anodic bond).
- <u>Die</u> bonding is the fixing of a sensor die on a board.
- <u>Wire</u> bonding makes the electric contact from the chip to the board.

The process of wire bonding is complicated, the bond loops need space and they are vulnerable and can be destroyed in a later process step. For this reason, many attempts have been made to replace the wire bonds by other means of making electric contacts. The most successful approach is flip chip mounting, Figure 3.15. To flip means to turn the chip upside down. One way would be soldering:

- Amplify the bond pads by screen printing or electroplating a solder
- Heat until the solder melts and, by surface tension of the liquid metal, contracts, to form a bump
- Flip the chip upside down and press it on the board
- Heat until the solder melts and connect chip and board (reflow soldering)
- To stabilize and protect the connection, let a drop of liquid glue flow under the chip and make it solidify there by UV-induced polymerization (underfill glue).

Flip chip technology is state of the art for electronic chips. Concerning microsystems, most of them are wire-bonded today, since the process of flip chip does not work for the vulnerable sensor structures such as membranes.

3.5.4 Integrating MEMS and Electronics

20 years ago, a pressure sensor was a sensor die in a TO socket, and it was connected to an external electronics. At present, most commercial devices

come as a system; sensor and electronics are integrated in one housing. There are strong reasons to integrate electronics:

- We get more integration within little volume.
- We save cost: only one housing, only one mounting process.
- Plug and play: When we mount a sensor in a car, it must work according to the specified characteristics. We cannot calibrate on the level of system integration, nor do we want to communicate calibration data of each sensor to the car. Therefore, the sensor must memorize its calibration data itself.
- Electromagnetic compatibility: Sensor voltage output is small, maybe only a few microvolts. A wire to the first preamplifier acts like an antenna, and electro smog will induce noise in the nanovolt or even lower microvolt range. This will ruin the signal-to-noise ratio. What we do is to put the first preamplifier on the chip or directly at his side. There, we amplify to the volt range, and microvolt stray voltage does not do harm any more.
- Reliability: A resistor or transistor normally does not fail unless overheated. Electronics fail at interfaces. Wire bonding is more robust than soldering, an aluminum interconnect on a chip is more robust than a wire bond. Therefore, we expect that higher level of integration will improve the long-term stability of our system.

Now we understand that we want to integrate electronics, but we have two options: monolithic integration means to build electronics and sensor on the same silicon chip. Hybrid integration means to make a sensor chip and an electronic ASIC separately, as shown in Figures 3.16 and 3.17. The trade-off decision is not easy, and it depends not only on technical considerations, but also on cost and sales numbers. It is much more difficult to develop monolithic sensors, and we need high sales numbers to reimburse development cost.

Arguments for hybrid integration:

- Less cost of development.
- Greater freedom of design. Monolithic integration restricts us to micromachining process steps, which do not harm the underlying electronics. This will not allow LPCVD anymore because of the high temperature.
- Separate technology lines with separate update of technology generations. ASIC lines make a generation update every 1–2 years. For the high number of chips they make, this pays off. Sensor lines make a generation

Figure 3.16 Hybrid integration: MEMS and IC are fabricated separately and mounted on the same PCB.

Figure 3.17 Monolithic integration: MEMS and IC are on the same silicon chip. In most cases, the IC is made first, with the MEMS on top of it.

update after 10 years. When I have developed a sensor, I want to produce it, and I do not want a change in technology.
- Yield: When I have to sort out a sensor for bad performance, I do not have to throw away the electronics with it.

Arguments for monolithic integration:

- Once developed, production can become cheaper at high numbers. The trade-off point seems to be quite high, at more than a million per year. This actually means automotive market and part of the consumer market.
- Small signals go directly to electronics. No parasitic capacity in the bond pads, no resistivity or inductivity in the bond wires. Very robust versus electromagnetic disturbances.
- Reliability: An aluminum line on a chip is more reliable than a bond wire.

Therefore, we expect monolithic devices at high sales numbers, in automotive, smartphone and part of the consumer market. However, there are also situations where monolithic integration is a must. Some examples are:

- The surface micromachined accelerometer must be monolithically integrated. The signal is a change in capacity, and the parasitic capacity of a bond pad would be much larger than this signal. By using bond wires before the first electronics, we would destroy the resolution.
- The multi-mirror device for projection with a beamer[8] has some millions of small mirrors, which are actuated 25 times per second. Let us assume 1 Mio. Pixels and 25 frames per second. Therefore, we need 25 MHz bandwidth for the signals. The information is sent in the chip at high frequency by about 80 bond wires. On chip, it is demultiplexed to construct 1 million signals of 25 Hz each. Without demultiplexing on chip, we would need a million of bond wires, which is not feasible.
- Medical devices such as the pressure sensor on a catheter to measure local pressure in a vein. If the whole system has less than 1 mm diameter, then we do not have space for hybrid integration of an ASIC.

In a recent research, wafer stacking becomes a major focus. An example from electronics is adding memory to a processor chip by stacking memory chips on it. Stacking can be done on wafer level or on chip level. In this way, the distinction between monolithic and hybrid integration becomes kind of obsolete. Nowadays, many people work on stacking MEMS to IC chips.[9] The functions are distributed on two pieces of silicon, which are bonded together vertically to form a wafer stack. The problem is how to get electric contact between the different floors. Figures 3.18 and 3.19 show possible strategies. The first uses metal eutectic bonding,[10] the second uses through-silicon vias. Figure 3.20 shows an accelerometer realized by wafer stacking.

3.5.5 Example of a Commercial Pressure Sensor

As an example for a state-of-the-art commercial pressure sensor, we can look at the MPX5100 sensor series by Motorola, which is well documented on the Motorola homepage.[11] It is a monolithically integrated sensor in a plastic housing, typically used for automotive or consumer applications.

The chip has a silicon membrane, but not the classic layout of four separate implanted regions. Instead, one resistor of window frame type is used,

---

[8]https://en.wikipedia.org/wiki/Digital_micromirror_device
[9]Integrating MEMS and ICs. A. C. Fischer, F. Forsberg, M. Lapisa, S. J. B.eiker, G. Stemme, N. Roxhed and F. Niklaus. Microsystems and Nanoengineering 1 (2015). https://www.nature.com/articles/micronano20155
[10]https://www.invensense.com/technology/
[11]https://www.ecse.rpi.edu/courses/CStudio/data%20sheets/DL200.pdf

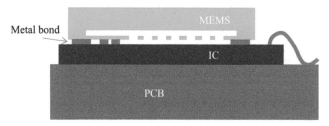

Figure 3.18 Wafer stacking. The MEMS is fabricated separately and bonded to the IC by metal bonding using the low melting point of an eutectic metal mixture. We still need bond wires.

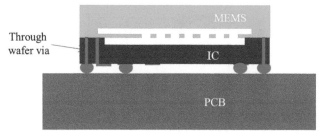

Figure 3.19 Another method for wafer stacking: The IC chip has vertical electric connections (through wafer vias). It is flip-chipped on the PCB board. No bond wires are needed.

Figure 3.20 An accelerometer made by wafer stacking.
Figure of R. Bosch GmbH.

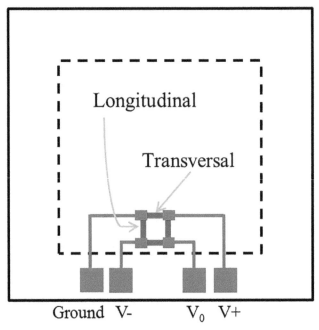

Ground V-   V₀ V+

Figure 3.21   A topical industrial pressure sensor of window frame type. The implanted area is a quadratic line. The four metal pads divide it into four resistors.

as shown in Figure 3.21. By the metal layer, this window frame resistor is divided into four resistors, two of them longitudinal and two transversal. This design is more robust versus technology inaccuracies, such as misalignment of the metal versus the local doping.

This chip is first mounted in an epoxy capsule, as shown in Figure 3.22. The capsule is made by injection molding, and electrical feedthroughs are embedded during molding, which is called the lead frame technology. The chip is die- and wire-bonded, and then it is covered by elastic silicone die coat to protect it against water, and the capsule is closed with a metal cover, which has a small opening, of course. Figure 3.23 shows a photograph of th epoxy capsule. For fluidic connection, there is a second housing, made by two shells with a fluidic port each. They are pressed around the capsule and joined by snap in place of plastic springs.

Plastic housing is the favorite technology for consumer and automotive sensors, as well as for electronic systems such as the common dual in-line package. It is cheap, but there is a major drawback: a plastic housing is never long-term-sealed against humidity. Water will diffuse through the plastic,

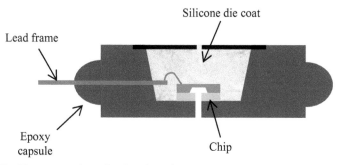

Figure 3.22 Epoxy capsule as first housing of a pressure sensor. Figure adapted from Motorola.[12]

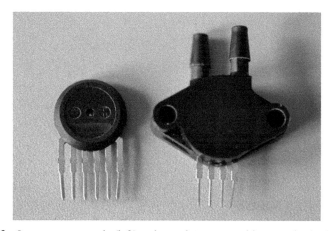

Figure 3.23 Inner epoxy capsule (left) and complete sensor with outer plastic shells (right).

which may cause failure. For this reason, we protect the silicon chip with a silicon nitride passivation layer and the bond wires with the silicone die coat. Alternatively, when we need to have long-term hermetic sealing, we could use a metal housing like the classic TO socket.

3.5.6 Sensor Electronics

For sensor readout, we use a classic Wheatstone bridge and an instrumentation amplifier. If you are already familiar with those, then please jump to the next chapter.

---

[12]https://www.ecse.rpi.edu/courses/CStudio/data%20sheets/DL200.pdf

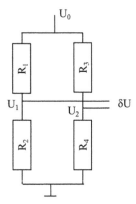

Figure 3.24 The Wheatstone bridge as readout for a piezoresistive pressure can compensate the temperature change of the resistors (TCR).

When pressure increases:
R_1 and R_4 decrease
R_2 and R_3 increase
Differential voltage δU increases.

When temperature increases:
R_1, R_2, R_3 and R_4 increase
Differential voltage δU is not affected.

The Wheatstone bridge shown in Figure 3.24 is the most important method to measure electric resistors. The resistors make two voltage dividers

$$U_1 = U_0 \frac{R_2}{R_1 + R_2} \quad U_2 = U_0 \frac{R_4}{R_3 + R_4} \tag{3.10}$$

Balanced bridge
We can change one of the resistors until the bridge voltage $U_1 - U_2$ vanishes. This is called a balanced bridge:

$$\delta U = U_1 - U_2 = 0$$

Then, $R_1 R_4 = R_2 R_3$
When the cross products are equal, the bridge is balanced.

Unbalanced bridge with one changeable resistor
When we do not balance the bridge, there will be a finite bridge voltage δU. First, we consider the case of one changeable resistor $R_3 = R + \delta R$ and three constant resistors $R_1 = R_2 = R_4 = R$. Then

$$\delta U = U_0 \left(\frac{1}{2} - \frac{R}{2R + \delta R} \right) = U_0 \frac{\delta R}{4R + 2\delta R} = \frac{1}{4} U_0 \frac{\delta R}{R} \qquad (3.11)$$

In the last step, we assume that $\delta R << R$. For strain gauges, this normally will be the case.

Four changeable resistors

Next, we consider the case of a pressure sensor with two resistors increasing and two decreasing:
$R_2 = R_3 = R + \delta R$; $R_1 = R_4 = R - \delta R$;
Then

$$\delta U = U_0 \frac{\delta R}{R} \qquad (3.12)$$

The main advantage of the bridge circuit is the temperature compensation. For a pressure signal, R_1 and R_4 are decreasing, while R_2 and R_3 are increasing, and a bridge voltage is generated. For increasing temperature, assuming a positive TCR of the resistors, all four of them are increasing. The voltages U_1 and U_2 do not change, and no bridge voltage is generated.

It is important to understand that this way the Wheatstone bridge alone can only compensate for temperature sensitivity caused by the TCR – the temperature coefficient of resistivity. This will reduce the temperature drift of the sensor, but not totally remove it. What remains is the temperature coefficient of the piezoresistive effect. For silicon, with increasing temperature, the piezoresistive coefficient k is reducing. Therefore, the sensitivity of the sensor will be reducing with increasing temperature.

Figure 3.25 shows a classic method to compensate for that by applying a series resistor R_S. This resistor must not be made from silicon, but from metal or carbon, in any case, it must have a much smaller TCR than the silicon resistors of the bridge. R_S and the bridge form a voltage divider. When temperature increases, the bridge circuit increases in resistivity. R_S also increases, but less. In this way, the supply of the bridge U_B increases, which will increase the bridge output. When R_S is properly trimmed, this increase in output cancels the decrease of piezoresistive constant k. Then, the sensitivity of the whole arrangement becomes independent of temperature.

Now we must find an electronic circuit to calculate the voltage difference $\delta U = U_1 - U_2$. We need to measure with high entrance impedance, since any measurement current would change Equation 3.10 and distort the measurement. To understand the instrumentation amplifier, we first reconsider operational amplifiers. Op-amps are differential amplifiers with a high

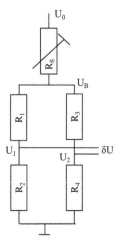

Figure 3.25   In silicon, the piezoresistive coefficient is a function of temperature. This cannot be compensated by the Wheatstone bridge alone. In this case, we can add a series resistor R_S made from different material with different TCR. Then, with increasing temperature, the bridge supply voltage U_B increases to compensate the decrease of piezoresistive coefficient.

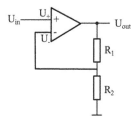

Figure 3.26   A noninverting amplifier using a negative feedback loop. It provides two advantages:
– High input resistivity
– No temperature problems since the amplification is only defined by the two resistors

internal gain k'. Typically, it is $k' = 10^6$. Figure 3.26 shows a differential amplifier with a negative feedback control loop.

$$U_+ = U_{in} ; \qquad U_- = U_{out} \frac{R_2}{R_1 + R_2}$$

$$U_{out} = k' (U_+ - U_-) = k' U_{in} - k' U_{out} \frac{R_2}{R_1 + R_2}$$

$$U_{in} - U_{out} \frac{R_2}{R_1 + R_2} = \frac{U_{out}}{k'} = 0 \qquad\qquad (3.13)$$

In the last step, my argument is that k' is very large, so $1/k'$ is extremely small and we can neglect all terms with k' in the denominator.

$$U_{out} = U_{in} \left(\frac{R_1}{R_2} + 1 \right) \tag{3.14}$$

This type of negative feedback amplifier is called noninverting amplifier or electrometer amplifier. It is very useful in measurement due to its decisive advantages:

- The amplification is only defined by the resistors, not by the op-amp. Temperature drift or long-term drift of the op-amp parameters have no influence.
- The electrometer has a very high input impedance and, in this way, measures voltages without measurement current.

The difference between the two inputs $\delta U = U_+ - U_-$ is very small. For an output of 1 V, it will be just 1 μV. We will neglect this and assume that the difference of U_+ and U_- is vanishing:

$$\delta U = U_+ - U_- = 0 \tag{3.15}$$

This assumption is called "ideal approach of the operational amplifier". In fact, the electrometer is a control circuit set to make $\delta U = 0$. To do this, it must generate a voltage output according to Equation 3.14. You may check whether this is a stable solution: Imagine that the output increases slightly. Then, the negative input increases, and the op-amp reacts in reducing the output until Equation 3.14 is fulfilled again.

Equation 3.15 is a general rule, which always holds when an op-amp is applied in a negative feedback loop. We will use this rule to understand the instrumentational amplifier in Figure 3.27.

For simplicity, we assume all resistors to be equal. Using Equation 3.15, we find two ways to calculate the current flowing through R_1:

$$I(R_1) = \frac{U_1 - U_2}{R} \quad \text{and} \quad I(R_1) = \frac{U'_1 - U'_2}{3R} \tag{3.16}$$

$$U'_1 - U'_2 = 3(U_1 - U_2) \tag{3.17}$$

U_+ and U_- are defined by two voltage dividers:

$$U_- = \frac{1}{2}(U_{out} + U'_1) \tag{3.18}$$

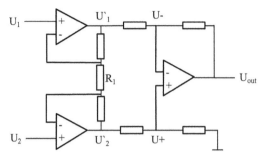

Figure 3.27   An instrumentational amplifier calculates the difference of two voltages and provides high input resistance. It is used for Wheatstone bridges.

$$U_+ = \frac{1}{2}U'_2 \qquad (3.19)$$

Using again the ideal approach of the op-amp, we request $U_+=U_-$ and obtain

$$U_{out} = U'_2 - U'_1 = -3(U_1 - U_2) \qquad (3.20)$$

The instrumentation amplifier provides a high entrance resistivity due to the two electrometers in the entrances, and it calculates the voltage difference. It is the most widely used electronics for Wheatstone bridges.

3.5.7 Calibration

As out of the clean room, a pressure sensor may provide an accuracy of $\pm 1\%$. We are not able to do better; deviations of membrane size and thickness and of doping of the piezoresistors will end up in a certain spread of process parameters. Many applications ask for better precision, $\pm 0.5\%$ or even $\pm 0.1\%$. For those, we have to do single-chip calibration, which means we have to measure the characteristics of every sensor and make the sensor memorize its specific data and to correct its measurement values accordingly. In the case of the pressure sensor, we may connect each device to a pressure source, measure at two or more different pressure values and then calculate the characteristics. This would allow us to take out the offset and to adjust the slope of the characteristics. But we must consider temperature: the temperature coefficient of sensitivity and offset must be adjusted to zero. This can be done by repeating the calibration in an oven by elevated temperature, which needs great effort. To avoid this, there is an alternative strategy: we can implement a heating resistor in the chip. We calibrate and then apply a heat pulse. We have to adjust parameters until the heat pulse will have no effect on the output any more.

How does the chip memorize its characteristic values? We need a non-volatile memory. The three most important technologies are:

1. Laser trimming on the hardware side: We apply resistors on the chip. Then, we cut out part of the resistors to increase their values. The disadvantage is that we must do trimming before we close the housing.
2. Fuses are devices that are irreversibly changed by a pulse of high current. A fuse is an interconnect with a constriction. The pulse will heat the fuse, and at the constriction, a hot spot is going to melt. An array of fuses combined with resistors allows us to trim the resistivity. There are also antifuses where a thin insulation layer is destroyed by a pulse. A Zener diode can be used as an antifuse by Zener-zapping, which is the destruction of the p–n junction by high temperature and electro migration. Fuses and antifuses are well known from the programming of FPGA (field programmable gate arrays). The advantage is that the calibration can be made after closing the housing. The disadvantage is that a strong current pulse must be generated on chip. This needs power transistors, which are normally not foreseen in sensor electronics.
3. EEPROM (electrically erasable programmable read-only memory) is a nonvolatile memory used in microelectronics for small amounts of data. It allows single bytes to be written and rewritten. This is elegant, but great effort in layout and technology is needed.

3.5.8 System Test, Reliability and Failure Modes

Finally, we have to test the systems. Test procedures are specified, especially in the automotive market. The essential idea is to accelerate the aging of a system by high temperature and to increase failure rate by overdoing negative impacts such as humidity. Failure rate is even more increased when voltage is applied during a stress test, which is called testing under bias. In this way, H3TB (high humidity, high temperature with bias) is a really nasty thing to do to a sensor system. We apply 80°C, 90% relative humidity for 1000 h. If this is not bad enough, then you can do temperature cycling: −40°C to +125°C; increase/decrease time = 5 min; retention time = 15 min; 1000 cycles. If your sensor survives, you have really done a good job as an engineer!

In microelectronics, it is common to estimate an expected lifetime from a failure rate via an accelerated aging test. You apply an Arrhenius approach, which describes the fact that many processes such as diffusion and chemical reaction proceed faster at elevated temperature. How much the rate k increases is determined by an activation energy E_a. R is the universal gas constant.

$$k = k_0 e^{\frac{-E_a}{RT}} \tag{3.21}$$

As a ballpark figure, we remember that a chemical reaction proceeds twice as fast for every a 10°C temperature increase. This approach is very useful if the failure mechanisms are based on diffusion and on chemical reaction. This is the case in microelectronics, where diffusion of dopants is a major source of failure. Also interface corrosion and water penetration in sensors follow an Arrhenius law. On the other hand, mechanical braking of membranes and many other sensor-specific failures will not follow an exponential law at all.

If sensors fail, how do they fail? It is difficult to get information from "out of the field". Some information is given, e.g. on the Motorola homepage. A dominant failure mechanism is corrosion of metals, such as interconnects on chip, bond wires, and lines on circuit boards. The danger is electrochemical corrosion: when two different metals touch, and when there is humidity, then there is a local galvanic element. An ionic current can flow and the less noble of the metals will dissolve. A well-known example is the connection of a copper rain pipe with a zinc one. Very soon the zinc, being less noble, will dissolve. Since we know that we will always have some humidity in plastic packages, there is always danger, e.g. the bond wires are gold, but the interconnects on the chip are aluminum or copper.

Silicon normally is stable, especially versus acids, but not at all versus bases (KOH is a base!). The much greater chemical danger is not on silicon, but on polymers, such as the silicone die coat in the pressure sensor. Plastic material may degrade chemically, especially if exposed to solvents or to fuel.

Fracture due to mechanical vibration is another source of trouble. Again, it is not the micromachined parts of a sensor, such as the spring mass system of an accelerometer, but it is the housing. Die and wire bonds may break when exposed to continuous vibration.

Electric failures are not common in pressure sensors, but in other MEMS where large voltages and currents are applied, such as actuators or capacitive sensors. An overview is given by Hartzell et al.[13] High currents can cause overheating, of course, but also electromigration. Moving electrons collide with atoms and they transfer momentum. If current density is very high, this can make the atoms move. Material is transported, which will weaken or even destroy interconnects. As a ballpark figure, electromigration becomes dangerous when the current density is 10^{10} A/m^2. If an interconnect is 5 μm wide and 200 nm high this means 10 mA, which is realistic in IC and in MEMS.

---

[13]A. L. Hartzell, M. G. de Silva, H. R. Shea: MEMS reliability. Springer 2011.

High voltages can destroy chips by dielectric charging or by electric breakdown. When we apply a potential of 30 V on an insulating layer of 300 nm, then the field strength will be 10^8 V/m. If there are pinholes or other impurities, this can generate electric breakdown and destroy the chip. But even if the insulation is faultless, at this high field strength some charges may move into the insulating layer by tunneling and diffusion. They get trapped in the insulating film, which causes a permanent loading. This is called dielectric charging. It modifies the characteristics of a sensor in an uncontrolled way, or it even causes irreversible blocking of movable capacitive structures of sensors and actuators.

Another danger of voltage is arcing, electric discharge in a gap between two structures. What voltage can be safely applied? As a ballpark figure, for a 4 μm gap and atmospheric pressure the breakdown will happen around 300 V to 400 V. An actuation voltage of 200 V over 2 μm gap is generally considered to be ok.

3.6 Capacitive Pressure Sensors

An alternative to piezoresistive transduction is capacitive sensing. We measure the deflection of a membrane by using it as one of the plates of a capacitor. When the membrane of area A is deflected down, the distance d of the plates reduces and capacity increases according to

$$C = \varepsilon_0 \varepsilon_r \frac{A}{d} \tag{3.22}$$

Capacitive pressure sensors have a number of advantages:

- Small-temperature cross-effect, far better than piezoresistive sensors.
- Small size: the chip is only 0.6 mm × 1.2 mm.
- No measurement current, low energy sensing possible.

On the other hand, capacitive pressure sensors are intrinsically nonlinear according to Equation 3.22, where the distance d is in the denominator.

Figures 3.28 and 3.29 show a capacitive pressure sensor in silicon technology. Small membranes of polysilicon are placed on a substrate of fused silica (highly purified amorphous glass). Why do we use glass as substrate and not the well-known silicon? The reason is that the substrate must be insulating. Otherwise, the metallization and the substrate would form a capacity parallel to the sensor capacity. This parasitic capacity would act as a short circuit for AC readout and would downgrade the capacitive sensor response. The whole sensor covers an area less than 1 mm^2. An advantage

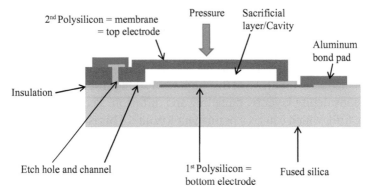

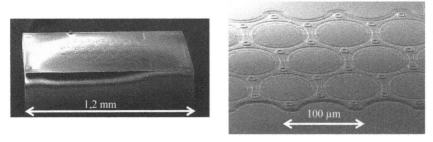

Figure 3.28   A capacitive pressure sensor in surface micromachining technology using a sacrificial layer for the generation of a cavity. Adapted from Schary et al.[14]

Figure 3.29   The capacitive sensor chip. Pressure sensors can be smaller than 1 mm. Each membrane has a diameter of 60 μm.
Chip by T. Schary, SEM by E. M. Meyer.

is the stability versus overload: in case of overload, the membrane does not break, but just touches down on the lower electrode as shown in Figure 3.30. For many applications, this touchdown mode is an important advantage compared to piezoresistive sensors, which are linear, but easily broken under overpressure.

This sensor is an example of a new technology approach called surface micromachining. Up to now, we have made mechanical structures in silicon by etching through the bulk of the wafer; this approach is called bulk micromachining. The problem is that we have to etch away 500 μm of a 525-μm-thick wafer just to get 25-μm-thick structures. This takes long time,

---

[14]Schary, T., M. Meiners, W. Lang and W. Benecke: Fused Silica as Substrate Material for Surface Micromachined Capacitive Pressure Sensors Operable in Touch-Mode. Proceedings of the IEEE Sensors 2005 Conference, Irvine, USA, 1–5 (2005).

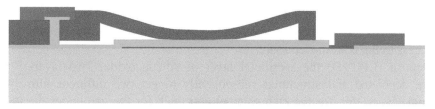

Figure 3.30 In overload, the membrane will touch down on the substrate. The characteristics will become flat, but the sensor is not going to brake.

Bulk micromachining   **Surface micromachining**

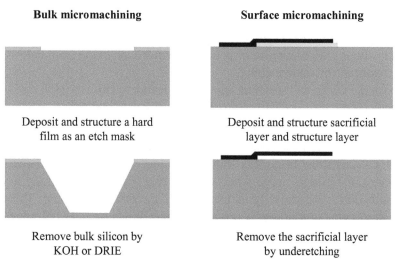

Deposit and structure a hard   Deposit and structure sacrificial
film as an etch mask   layer and structure layer

Remove bulk silicon by   Remove the sacrificial layer
KOH or DRIE   by underetching

Figure 3.31 Comparing bulk and surface micromachining.

and precision is limited. An alternative would be to do everything from the top, and this is surface micromachining. But how can we get a cavity to have a free-standing membrane or a movable part? This can be done with a sacrificial layer. Figure 3.31 shows the basic idea. As a sacrificial layer, we deposit and structure silicon oxide. Then follows the structure layer made of polysilicon. Now we remove the sacrificial layer, which can be done by wet chemical etching in HF. HF etches oxide, but not silicon. We need good selectivity because we have to underetch the polysilicon structures without harming them. Surface micromachining has been developed in the 1990s; today it is used mainly for accelerometers and gyroscopes, but also for pressure sensors. The structures are very small and precise, which is an advantage for capacitive sensing.

For making a capacitive pressure sensor as in Figures 3.28 and 3.29, the important technology steps are the following:

1. Deposit 1st polysilicon (LPCVD) for the lower electrode and structure it.
2. Deposit (LPCVD) silicon nitride insulation and structure it for electric access to the bottom electrode.
3. Deposit (LPCVD) the sacrificial layer of silicon oxide. Two layers are deposited and structured subsequently to get two different film thicknesses: thin for the etch channel and thick for the gap under the membrane.
4. Deposit 2nd polysilicon (LPCVD) for the membranes and structure it to open the etch hole.
5. Remove the sacrificial layer by etching the silicon oxide by HF through the etch hole and the etch channel. HF etches silicon oxide, but does not attack silicon. Allow long etching time to remove the oxide from below the membrane by underetching.
6. Deposit (LPCVD) and structure another oxide film to close the etch hole.
7. Deposit (sputtering) and structure aluminum for the bond pads.

This device will give us a pressure signal in the form of a small change in capacity. How will we make electronic readout? To measure capacity, we could use a Wheatstone bride fed with alternating current. But to realize its advantages, the bridge circuit wants to have two transducers working in opposite direction – one capacity increasing, and the other decreasing. This is not the case here, so a bridge is not appropriate. When there is only one capacity, a frequency analogous readout is the better choice.[15]

We construct an astable multivibrator whose frequency is determined by the capacity C together with a resistor R, as shown in Figure 3.32. We use an operational amplifier as a switch ("comparator mode"). Let us assume we start with a positive output voltage $U_a = U_0$, which is the supply voltage of the op-amp. The positive input port has a voltage of $U_+ = 1/2\ U_0$. The capacity is loaded via the resistor R and U_C is increasing. When U_C outbalances U_+, the op-amp is going to switch and the output changes to $-U_0$ and $U_+ = -1/2\ U_0$. Now the capacity is unloaded until U_C outbalances U_+ again and the op-amp switches again. We get a rectangular output function, and for a rough estimate, we assume a period time of twice the RC time constant, one for loading and one for unloading.

$$f \approx \frac{1}{2RC} \tag{3.23}$$

---

[15]Elmar Schrüfer: Elektrische Messtechnik, Hanser Verlag.

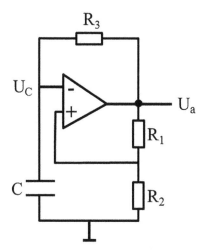

Figure 3.32 An astable multivibrator for capacity-to-frequency conversion.

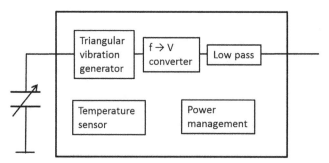

Figure 3.33 A monolithic capacity measurement unit by analog microelectronics as used today commercially.[16]

The advantage of a frequency analogous readout is that it can be evaluated easily by counting the rectangular function, which is very robust with respect to noise.

For very sensitive capacity measurements, monolithically integrated converters are available, e.g. the CAV444 shown in Figure 3.33. The first step is a capacity-to-frequency converter. The capacity to be measured is charged and discharged periodically with constant current, thereby generating a triangular vibration. Then, a frequency-to-voltage converter is used to generate a voltage output.

---

[16]https://www.analog-micro.com/de/produkte/ics/cu-wandler/cav444/

3.7 The Recent Trend: Additive Technology and Printed Sensors

Printing has been since long a method to make metallic interconnects on circuit boards. In the last years, methods have been improved, and printed layers are not only interconnects any more, but also functional layers for sensors. Printing does not need a clean room and expensive equipment like sputtering or lithography. Using 3D printing, even complicated three-dimensional forms can be generated. Fully printed sensors may replace silicon sensors in future for many applications. The key challenge is the long-term stability.

When we print, we add material on a substrate as an ink or a paste. Screen printing (Figure 3.34) Screen printing is a mature and established technology for circuit boards. We take a screen of steel wires with a typical mesh size of 50 μm. The screen is soaked in resist and dried. Then, we carry out photolithography and develop the resist. Where we want the paste to deposit, the meshes of the screen are open.

When the screen is prepared, we can start printing. We lay the screen on the substrate. We place some paste on the screen and we squeeze the paste through the holes on the substrate using a rubber spatula called squeegee. We remove the screen and anneal at elevated temperature. When printing polymer, we can anneal at 100°C. When we want to produce metals with good properties, annealing becomes critical. We can get a conductive film when we anneal at low temperature such as 90°C, but the film will not be stable. We obtain good metallic films only when the particles in the film grow together by diffusion, which is called sintering. But for diffusion, we need to go to zone 2, better zone 3 in the three-zone model of thin films (Chapter 2, Figure 2.5), and we end up using up to 800°C for film sintering.

For interconnects, silver particles embedded in a solvent is the most common ink. Other pastes are made on the basis of a liquid polymer such

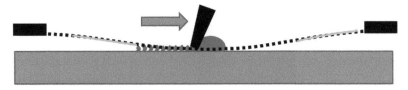

Figure 3.34   Screen printing: a paste is squeezed onto the substrate through a screen. In this way, the structure in the screen is transferred. Then, the paste is dried and annealed to get a continuous film of some micrometer thickness.

as PDMS (polydimethylsiloxane) with embedded fillers such as carbon particles, carbon nanotubes or graphene. Also, semiconductors may be printed, e.g. ZnO. For insulating films, we can use epoxy or PDMS.

Important applications of screen printing are circuit boards, organic light-emitting diodes (OLEDs), thin-film transistors and sensor layers (piezoresistive films, gas sensing films and others).

Alternative printing technologies are too many to describe, and I want to briefly list some with increasing importance[17]:

- Inkjet printing: we apply small drops of ink dispersed from a nozzle. Resolution is 20–50 μm, but the process is slow when compared to screen printing.
- Aerosol jet printing: very small droplets (5 μm) are driven by an air stream. The advantage is the high speed of the droplets, which generates high impact of the droplets hitting the substrate and, in this way, high film quality. Resolution can be as small as 10 μm, and films as thin as 100 nm can be made.
- Imprinting: we use a structured stamp to deposit a liquid film. Imprinting technology became important as nanoimprinting, which has better resolution than optical lithography and allows structures in the nanometer range.

Why do we want to print sensors and electronics when a mature technology is available using silicon technology? Printing can be much cheaper, and it is a simpler technology that has less process steps. On the contrary, quality of the devices is not as good as in silicon technology and high integration density cannot be achieved.[18] In Table 3.2, three technologies for making pressure sensors are compared: silicon technology, thin film on steel and printing.

Figures 3.35 and 3.36 show examples of printed pressure sensors. In Figure 3.35, a piezoresistive sensor layer is printed on a foil substrate. There is no sensor membrane, but the compression of the material itself is measured. This type of pressure sensors can be used for material integration. Examples are sensors in gaskets or sensors in fiber compound material.

---

[17]Technologies for Printing Sensors and Electronics Over Large Flexible Substrates: A Review. S. Khan, L. Lorenzelli, R. S. Dahiya. IEEE Sensors Journal, 2015.

[18]Sabine Globisch: Lehrbuch Mikrotechnologie, Hanser ed.

Table 3.2   Comparing technologies for pressure sensors

	Silicon Pressure Sensor	Thin Film on Steel	Printed Sensor
Substrate	Silicon wafer	Steel	Foil
Piezoresistive material	Monocrystalline Si	Sputtered metal	Polymer with carbon particles
Insulating material	LPCVD	PECVD	Printed polymer
Structuring	Photolithography	Photolithography	Screen printing
Maximum process temperature	1000°C	350°C	100°C
Number of process steps	18	7	3
Size	2 mm	5 mm	Some mm
Prize	Cheap at large numbers	Expensive	Potentially very cheap
Stability	Good	Very good	Not yet known
Integration in material	As chip by hybrid integration	As system	Foil integration direct in the matrix material
Technology	Mature	Mature	Experimental status

Sensor layer

Electrodes

Foil substrate

Figure 3.35   A printed pressure sensor.[19]

Figure 3.36 shows an experimental device of a fully printed pressure sensor of the membrane type. The corpus including the membrane has been made by 3D printing, the piezoresistors are added by screen printing. Maybe we will be able to fabricate fully printed force or pressure sensors in future. The crucial question is the long-term stability of the 3D-printed mechanical parts.

The corpus of the sensor in Figure 3.36 is made by 3D printing. This technology is gaining more and more interest, also for microsystems. Structures with lateral definition of approximately 10 μm, which are made

[19]Gräbner, D.; Tintelott, M.; Dumstorff, G.; Lang, W. Low-Cost Thin and Flexible Screen-Printed Pressure Sensor. Proc. Eurosensors, Paris, Frankreich, p 616, 2017. DOI 10.3390/proceedings1040616.

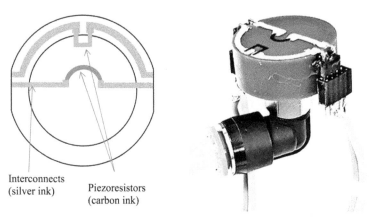

Interconnects
(silver ink)   Piezoresistors
                  (carbon ink)

Figure 3.36 Fully printed pressure sensor. Layout (left) and final device (right). Figure with friendly permission of J. Hinz and W. Gehlken.[20]

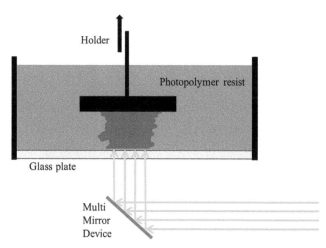

Figure 3.37 A 3D printing system using stereolithography. When the liquid resist is illuminated, it solidifies. An object can be written layer by layer out of the liquid resist.

by bulk silicon micromachining today, may be done by 3D printing in the near future. The principle is shown in Figure 3.37. The object is placed on a movable platform in a tank of liquid resist. The resist contains the monomers of a photopolymer, e.g. epoxy resist. There is a thin layer of liquid resist between the object and the glass bottom of the tank. Where we want the

---

[20]J. Hinz, W. Gehlken, F. Lucklum, G. Dumstorff, W. Lang and M. Vellekoop: Additive Fertigung mechanischer Sensoren. Sensor&Test, Nürnberg (2018).

resist to harden, we illuminate it. The illuminated liquid will polymerize and solidify. Then, we lift the holder by one layer thickness. In this way, layer by layer, a 3D object is written. Illumination can be done by a moving laser focus or by a multimirror array like in a beamer. The thickness of the working layer and lateral resolution are around 100 µm for macroscopic systems. There are also high-resolution systems with layer thickness and lateral resolution of some micrometers.

An example for a part that has been made in silicon and is now made by 3D printing is the gas chromatography column shown in Figure 3.38. A gas chromatography column is a small and long tube filled with an adsorbing powder. When a gas mixture is given to the column, due to different times for adsorption and desorption, the components are separated and gas composition can be analyzed. The classic way to make these columns is steel or glass tubes. Some years ago, microfluidic devices etched in silicon have been published for this purpose.[21] In 2015, a printed column has been presented, as shown in Figure 3.38. The worries were that the plastic might outgas

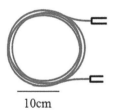

10cm

| Column from a thin metal tube as it is used today | A silicon column made of a triple stack of Pyrex, silicon and Pyrex. The structuring is done by DRIE etching[22]. Side length 25 mm. | Gas chromatography column made by 3D printing[23]. Side length is 35 mm. The resolution for ethylene is in the lower ppb range, as good as for the silicon devices. |

Figure 3.38   Development of Gas Chromatography Columns by MEMS technology.

[21]Real-time monitoring of sub-ppb concentrations of aromatic volatiles with a MEMS-enabled miniaturized gas-chromatograph. S. Zampolli, I. Elmi, F. Mancarella, P. Betti, E. Dalcanale, G.C. Cardinali, M. Severi. Sensors and Actuators B 141 (2009).

[22]Stürmann, J., W. Benecke, S. Zampolli, I. Elmi, G. C. Cardinali and W. Lang: A micromachined gas chromatographic column to optimize the gas selectivity for a resistive thin film gas sensor. Transducers'05, Seoul, Korea (2005).

[23]Zaidi, N. A.; Tahir, M. W.; Papireddy Vinayaka, P.; Lucklum, F.; Vellekoop, M. J.; Lang, W. Detection of Ethylene using Gas Chromatographic System. Eurosensors (2016).

and this way inhibits high-resolution measurement, but this is not the case. The column shown in Figure 3.38 was used to measure ethylene in fruit transport, and a resolution in the ppb (parts per billion) range has been demonstrated.[24]

Figure 3.39 shows the focus of a laser. The photons used for photopolymerization have a wavelength of $\lambda = 365$ nm, which restricts resolution to some hundred nanometers. However, there is a method to write even smaller structures: two-photon processes. We apply near-infrared light at $\lambda = 780$ nm. These photons cannot induce chemical reaction, they do not have enough energy. But sometimes, two photons are absorbed by the same molecule in the same moment, then the energy adds. The two photons act like one photon with 390 nm wavelength and a chemical reaction will happen. To increase the chance of two photon processes a laser with short but strong light pulses is used. Still the chance of a two-photon process is only given in the center of the focus, where the intensity is highest. This small volume with highest energy (red in the figure) is only 100 nm wide; in this way, the two-photon process focus is decisively smaller than the normal optical focus. Using this, we can write structures with 100 nm lateral resolution. The two-photon configuration has a second big advantage: there are resists that are transparent for the

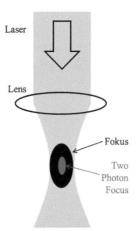

Figure 3.39 Two-photon lithography. Visible light can realize a focus of some hundred nanometers. Two-photon processes have a smaller focus of 100 nm.

---

[24]Lucklum F., S. Janssen, W. Lang, M.J. Vellekoop: Miniature 3D gas chromatography columns with integrated fluidic connectors using high-resolution stereolithography fabrication. Eurosensors (2015).

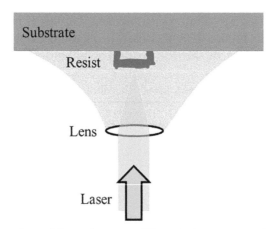

Figure 3.40   Two-photon lithography system. The resist is transparent for the laser light, so we can write directly in the resist.

infrared light of the laser. In this way, we can directly write in the film, as shown in Figure 3.40.

Figures 3.41 and 3.42 give examples of microstructure via two-photon lithography. Figure 3.41 shows optical waveguides written over a silicon wafer. The idea is to use light for transporting data on chip in future data-processing systems. Figure 3.42 is a test structure to show that lateral resolution of 100 nm can be reached. Two-photon systems can do things that

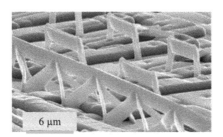

Figure 3.41   Microoptic structures written by two-photon lithography using a nanoscribe system.[25] You see an optic waveguide written over the surface of a silicon wafer.[26] Figure by courtesy of BIAS, Bremen, Prof. R. Bergmann.

---

[25]https://www.nanoscribe.de/de/

[26]M. Schröder, M. Bülters, C. von Kopylow, R. Bergmann: Novel concept for three-dimensional polymer waveguides for optical on-chip interconnects. Europ. Opt. Soc. Rap. Public. 7, 12,027 (2012).

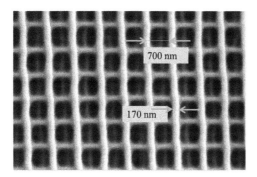

Figure 3.42 A test structure written by two-photon lithography to show lateral resolution of 100 nm.
Figure by Frieder Lucklum, IMSAS.

classic microtechnology cannot achieve: they realize real 3D structures, and they can make them as small as 100 nm.

Questions

- Explain the piezoresistive effect. Derive the piezoresistive formula 3.2 and explain it.
- Why is the piezoresistive coefficient k in metals around 2, but not exactly equal to 2.0?
- Why is k in silicon larger than in copper?
- Draw the geometry of the resistors on a pressure sensor membrane of steel.
- How does a piezoresistive sensor in silicon technology work?
- Draw the geometry of the resistors on a pressure sensor membrane of silicon.
- How does a capacitive pressure sensor work?
- What is a sacrificial layer?
- Draw and explain a 3D printer
- Explain a noninverting feedback amplifier using an op-amp.
- Draw and explain an instrumentation amplifier
- What is flip chip bonding?
- What are the pros and cons of hybrid and monolithic integration?
- How do you measure via resistance?
- Explain screen printing.
- How does wire bonding work?
- Explain the process of 3-D printing

4

Inertial Sensors

4.1 Inertial Measurement of Motion

For measuring motion, there are two fundamentally different strategies: The first is measurement by direct contact to a fixed reference frame. An example is the speedometer of a car. We use the road as a reference, and we assume that there is no slip between the tire and the road. Then, we count the rotations of the tire and we can calculate the distance traveled and the speed we have. But what if there is slip? The road is icy, I brake and the wheel does not turn any more. Then, the speedometer tells me I am standing, while I am slipping on.

In this case, the second possibility will help: inertial measurement. When a mass is accelerated or decelerated, there must be a force causing the change of motion, which is called the Newton force. I can measure this force, and, integrating acceleration over time, I know the speed. The advantage is that no external reference is needed. The disadvantage is that speed can only be determined by integration, but not by direct measurement.

What can be measured inertial (Table 4.1)? Acceleration yes, but speed not. Newton himself describes a very interesting thought, as shown in Figure 4.1. As a simple inertial sensor, you got a bucket half-full of water. When you move it with a constant speed, the surface of the water stays flat, and you see nothing. It is not possible to measure constant speed inertial. When you accelerate, the surface will tilt, thus linear acceleration can be measured inertial.

What about rotation? When we rotate the bucket, the water is driven to the outside by the centrifugal force, and the water surface will get curved. Constant rotation generates a force, and a sensor for angular rate can be made, also known as micro gyroscope. In fact, rotation generates two inertial forces: centrifugal force and Coriolis force, and when we discuss micro gyros, we will find the latter much more appropriate for inertial sensing.

143

Table 4.1   What can be measured by inertial sensors?

Linear Motion			Rotation		
Location	x	No	Angle	Φ	No
Velocity	$v = dx/dt$	No	Angular velocity	$\Omega = d\Phi/dt$	Yes
Acceleration	$a = dv/dt = d^2x/dt^2$	Yes	Angular acceleration	$d\Omega/dt$	Yes

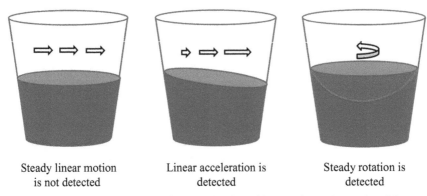

| Steady linear motion is not detected | Linear acceleration is detected | Steady rotation is detected |

Figure 4.1   Newton's bucket experiment. A bucket with water is used as an inertial sensor. What type of motion can we detect?

4.2 Acceleration Sensors

4.2.1 Applications and Ranges

Acceleration a is the derivative of the speed of movement v. It causes an inertial force F, the Newton force, which is proportional to the inertial mass m.

$$\vec{a} = \frac{d\vec{v}}{dt} = \frac{d^2\vec{x}}{dt^2} \tag{4.1}$$

$$\vec{F} = m\vec{a} \tag{4.2}$$

The dimension of acceleration is 1 m/s^2. In sensors, the average acceleration of gravity is the preferred unity: 1 g = 9.8 m/s^2.

Applications and ranges for acceleration measurement are:

• The most important application is the airbag which triggers around 50g. To prevent triggering of the airbag by false alarm, airbag systems have two sensors: (1) an acceleration sensor and (2) a mechanical acceleration switch. Both must be positive to trigger the airbag. While first airbag systems just had a trigger point, modern systems analyze the acceleration–time signal. In this way, a real incident can be distinguished from other harmless acceleration, such as driving over the curb too fast.

- Anti-blocking systems (ABS, also anti-skid system or anti-lock braking system) for the brakes of a car. When the road is icy and you brake, the wheels may block. With blocking wheel, you cannot steer any more, the car just slides on. The anti-blocking system measures the rotation of the wheels and the acceleration. When the wheels do not rotate any more, but the car fails to decelerate, the system takes action. The pressure of the brakes is reduced for a short moment, allowing some rotation. In this way, the contact of the tire and the road is not lost, more braking power can be generated and some steering will be possible.
- In active-suspension of a car, the vertical distance of the wheels with respect to the chassis is controlled using an actuator. The idea is to reduce bumping of the car on a rough road. This increases comfort, as well as security. Better contact of wheel and road allow better steering and braking. The control is done with an accelerometer for vertical movement of the chassis. The system operates the actuators in a way to minimize this vertical movement.
- Active vibration damping: wherever you want to suppress a vibration, active instability control can be applied. You measure the vibration with an accelerometer. Then, you mount an actuator to the system and counteract the vibration. This can be used for engine vibrations, structure vibrations, sound and many others. An electron microscope, for example, must be placed on a vibration free bench to reach its full resolution. The same holds for many other scientific instruments. Classically, this is done by a spring mass system with extremely high mass, such as tons of concrete. A new approach is active vibration isolation. You measure the vibration of the table with six accelerometers. Using actuators as table supports, you control the vibration to zero.[1]
- Many accelerometers are used to measure the movement of persons or of items in their surroundings. In gaming, man–machine interaction is achieved by game stations equipped with accelerometers.
- Laptops have accelerometers as "free fall sensors" for the security of their hard disk drives. When you drop the labtop, the acceleration suddenly falls from 1g to 0. Then, the reading head is quickly removed

---

[1] http://www.mciair-opticaltables.eu/product/ta600-active-vibration-isolation-platform/.

from the hard disk. In this way, when the laptop finally hits the ground, a deadly crash of the reading head and the rotating disk is prevented.

- You can mount an accelerometer on a body and measure structure vibrations. For body sound, an accelerometer is an alternative of a microphone. When a wheel bearing in a car is worn, the sound changes. This can be used for health monitoring. When moving parts start to degrade by mechanical abrasion, their movement becomes rough and noise increases. Imagine a large gearbox in an offshore wind energy plant. There, it is essential to detect a defective toothed wheel before it breaks. Analyzing the vibrations of the structure can warn you before a catastrophic event. It is even possible to analyze which of the many toothed wheels in the gearbox has a problem, since they rotate with different rotation speeds.

- Figure 4.2 shows the "bathtub curve", a general feature of failure versus lifetime shown by many technical systems. Initially, failure rate is high. Early failures are hidden problems in production, which appear after a short period of use. Then, for the length of normal lifetime, failure rate is constantly low. When the systems reach their lifespan, due to wearout failures, the curve rises again. The interesting thing is the vertical axis. It can be failure rate, but when we measure vibration amplitude, body sound or sound emission, we find very similar curves: New systems run roughly, and after some time, working becomes smooth. Finally, parts wear out, and movement roughens again. Increase of vibration and noise, in many cases, indicates approaching breakdown.

- Inclination measurement is a special case of acceleration measurement. Here, we do not want to know the amount of acceleration, but the direction of the vector. This can be done by a 2-axis accelerometer, e.g.

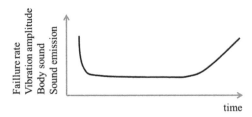

Figure 4.2   The bathtub curve. At the beginning and at the end of the lifetime, there is a high failure rate and a high level of vibration and noise. In between, we find a long range of smooth working.

for the rotation of the screen in a smartphone. For security applications, inclination has to be measured better than 1°, and an example is to control the vertical position of a large derrick when it lifts a heavy weight. For that, accelerometers are not sensitive enough, and there are special inclination sensors using pendulums or liquid surfaces.

4.2.2 Sensor Principles and Resonance

To measure the acceleration, we can use a spring mass system, as shown in Figure 4.3. The important parts are:

- A seismic mass, which generates a Newton force when accelerated.
- A spring with spring constant k, which transduces this force in a mea-surable extension. In this way, measurement of acceleration is reduced to the measurement of elastic deformation, which is known from force sensors.
- We have to measure the elastic deformation by measuring the movement amplitude of the seismic mass relative to the accelerometer $x(t)$.
- A spring mass system has a time behavior of degree 2, which means it can overshoot. To prevent this, we need a damping b.

To understand the dynamics, we look at the resonance curve of a spring mass system (forced and damped harmonic motions). Figure 4.4 plots the vibration amplitude of the seismic mass versus the frequency when it is subject to a sinusoidal external movement.

We start with the force balance. The mass is subject to three forces: (1) inertial force, (2) damping force and (3) elastic force of the spring. These

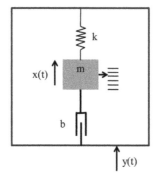

Figure 4.3 Essentially, an accelerometer is a spring mass system in a closed box. Mass m, spring constant k, damping b. The movement of the box as a whole is described by $y(t)$. The internal movement of the mass with respect to the box is $x(t)$.

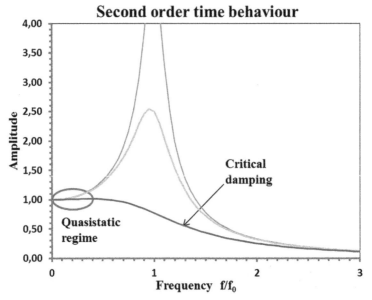

Figure 4.4   Resonance curve of a spring mass system. For an accelerometer, we prefer critical damping since it will prevent overshooting. We want the resonance curve to be flat; this restricts us to small frequencies with respect to the resonance frequency. This is called the quasistatic regime.

balance the external driving force (4):

$$m\frac{d^2x}{dt^2} + b\frac{dx}{dt} + kx = F_{ext}e^{i\omega t} = ma_{ext}e^{i\omega t} \qquad (4.3)$$
$$\quad (1) \qquad (2) \quad (3) \qquad (4)$$

Using the ansatz $x = x_0 e^{i\omega t}e^{i\varphi}$, we can find the amplitude:

$$\frac{dx}{dt} = i\omega x \text{ and } \frac{d^2x}{dt^2} = -\omega^2 x$$

We substitute this into the differential equation and get

$$\left(-m\omega^2 + ib\omega + k\right)x_0 e^{i\omega t}e^{i\varphi} = ma_{ext}e^{i\omega t}$$
$$\left(m\omega_0^2 - m\omega^2 + ib\omega\right)x_0 e^{i\varphi} = ma_{ext}$$

with the resonance frequency $\omega_0 = \sqrt{\dfrac{k}{m}}$ \qquad (4.4)

For the amplitude we find:

$$\left(m^2 \left(\omega_0^2 - \omega^2\right)^2 + b^2\omega^2\right) x_0^2 = m^2 a_{ext}^2$$

$$x_0 = \frac{a_{ext}m}{\sqrt{m^2\left(\omega_0^2 - \omega^2\right)^2 + b^2\omega^2}} \tag{4.5}$$

Around the resonance frequency, we find a resonance magnification if the damping is low. Low damping and resonance magnification would have the consequence that, when exposed to a step response, the sensor would overshoot, as shown in Figure 4.5. This is not acceptable and therefore we apply more damping. If we do too much damping, then the movement of the mass will be seriously inhibited, and the step response becomes very slow, which reduces bandwidth. The best value is critical damping, which is the case for

$$b = 2\sqrt{mk} \tag{4.6}$$

or Q = 0.5, using the quality factor of a resonator $\quad Q = \dfrac{\omega_0 m}{b}. \tag{4.7}$

For low frequencies, the curve is flat at an amplitude of 1. This is called the quasistatic regime. For the accelerometer, the resonance curve must be flat. Otherwise, the sensitivity of the accelerometer would be a function of the frequency, but we want the same sensitivity for all signal frequencies.

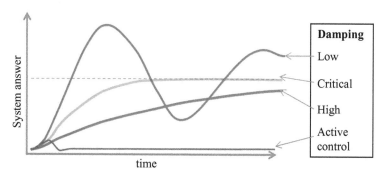

Figure 4.5 Step response of a second-order time behavior system. For low damping, the system will overshoot. For very high damping, the system will approach the final value very slowly ("creeping solution"). The ideal value is the critical damping.
A different approach is active control: we do not allow the mass to move, but we actively keep it in place. The acceleration is then measured by the force we need. We do not have to wait for equilibrium, and therefore, the active approach can be very fast.

This is only the case in the quasistatic regime. In this case, Equation (4.5) is reduced to:

$$\omega \ll \omega_0 \rightarrow x_0 = \frac{a_{ext}}{\omega_0^2} \tag{4.8}$$

The mass vanishes in Equation 4.8. At first sight, this is astonishing. When I want to build a sensitive accelerometer, I would use a large mass. Large mass means large Newton force, and also large movement of the spring mass system and large signal. Wrong! We presume that the resonance frequency is fix, because it is normally given by the application. Now, when mass doubles, we have to double the spring constant to keep resonance frequency fixed. Thus, the movement does not increase nor does the sensitivity. In fact, accelerometers in silicon technology have extremely small masses in the microgram range.

For a practical example, we look at the airbag accelerometer. We have 3 ms to do measurement of acceleration, analysis of the movement and firing the airbag. For the sensor, we can allow a maximum response time of 0.25 ms or a minimum bandwidth of 4000 Hz. The quasistatic regime must be zero to 4000 Hz, then Figure 4.4. shows that resonance frequency must be 10 kHz, not less. In most application cases, there will be a specification about minimum bandwidth.

When increasing mass does not work, what can we do to increase accelerometer sensitivity? Equation 4.8 shows us the way: we have to measure x, the displacement of the mass, as good as possible. The resolution of measurement of displacement, δx, is the limiting factor for the resolution of the accelerometer.

4.2.3 Layout and Technology

An accelerometer is a spring mass system in a box, plus a sensor to measure the displacement of the mass, and the box will be the silicon chip. The measurement of the displacement is critical, since it determines the resolution of the sensor. We know several methods to measure displacement: piezoresistive, piezoelectric and capacitive. All methods have been tested in different types of accelerometers, and capacitive sensing did clearly win. Using surface micromachining and comb structures for measurement capacities, we can resolve relative displacements smaller than a nanometer.

The principles of surface micromachining we have already discussed for the capacitive pressure sensor. We need a sacrificial layer and a structure layer. The structure layer should be thick, to increase capacity of the

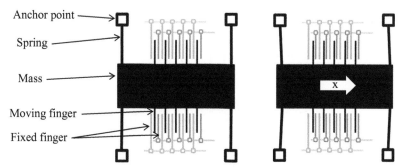

Figure 4.6 Principal geometry of a surface micromachined accelerometer. Top view of the structure layer. When the mass moves due to acceleration, the capacities between fixed fingers and moving fingers change. Note the differential capacity: for each moving finger, there are two fixed fingers on different potential.

comb structures. Today, typically a polysilicon of 4 µm to 10 µm thickness is used. As a sacrificial layer, silicon oxide is the ideal partner for polysilicon. Both layers are made by high temperature oven processes, and selective etching is easily possible using hydrofluoric acid HF, which will dissolve oxide but not attack the silicon.

The challenge of the layout is to put all functions in one structure layer: mass, springs and comb sensors. Figure 4.6 shows a schematic. The mass is fixed with four springs, which are anchored on the chip at the other end. When the mass can move in x-direction only, it is a 1-D accelerometer. The mass has a row of comb fingers, which move with the mass. For each moving finger, we have two fixed counter fingers, forming a differential capacity. Note that these two counter fingers must be on different potentials.

When the whole system is accelerated to the left, Newton force on the mass is to the right (observing in the reference system of the sensor). The springs are elongated until the elastic force cancels Newton's force.

Figure 4.7 shows a two-axial accelerometer, as it is used today. The structure layer is made of thick polysilicon. The seismic mass is held by four meandering springs, which allow movement in x and y directions. For easier removal of the sacrificial oxide, the seismic mass has etch holes, which are not drawn in the figure. We can see the moving electrodes at all four sides (for biaxial measurement) with two fixed electrodes for each moving one. The fixed electrodes are made in the thick polysilicon of the structure layer, and each of them has an anchor, which are connected to two buried electrically separate interconnects.

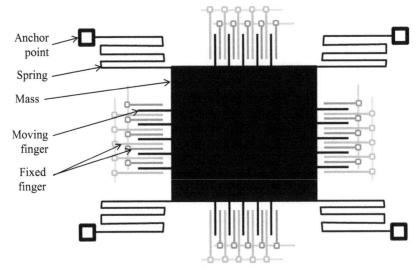

Figure 4.7   Schematic of a two-dimensional accelerometer.

From a picture of a biaxial accelerometer by Analog Devices,[2] we can estimate the seismic mass to be about only 1 μg. A corn of rice weighs about 10 mg, so we would need 10.000 of these seismic masses to outweigh a corn of rice. This impressively demonstrates Equation 4.8: mass is not needed for sensitivity, but precise detection of δx. This is done by very precise etching of the capacitive fingers. The etched gap is 2 μm, which allows a resolution of δx in the sub-nanometer range.

How is the accelerometer damped to prevent overshooting? Fortunately, we do not have to do anything but leaving a little gas in the system. Viscous damping in the gas will be enough to get critical damping.

Thick polysilicon is obviously of advantage since it gives us higher capacities. An improved LPCVD process provides polysilicon with extremely low stress, which is used to make free-standing cantilevers thicker than 10 μm. This material is called epi-poly silicon. Figures 4.8 and 4.9 show surface micromachined accelerometers made by R. Bosch GmbH. In Figure 4.8, you can see the configuration of the measurement fingers: one moving finger is opposed by two fixed fingers. These must be on two different potentials. For this reason, one line is electrically connected in the structure layer. The other has an anchor for each electrode finger. They are connected in a buried layer below.

---

[2]https://www.indiamart.com/proddetail/mems-accelerometers-10275047112.html.

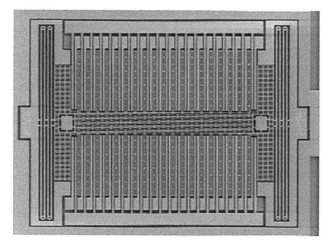

Figure 4.8 A silicon accelerometer chip made by R. Bosch GmbH.
You can see the folded springs holding the seismic mass with the moving fingers. Figure by courtesy of R. Bosch GmbH, Dr. F. Laermer.

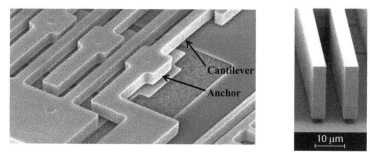

Figure 4.9 Technology details of an accelerometer. Left: Free-standing cantilevers after removal of the sacrificial layer. You can see the anchors, which are not underetched.
Right: Cantilevers made of thick and stress-free polysilicon (epi-poly), after DRIE etching and sacrificial layer removal. Figures by courtesy of R. Bosch GmbH.

4.2.4 Accelerometer Electronics

The movement of the mass in the accelerometer is extremely small. From Equation 4.8, we can calculate the movement amplitude. For an acceleration of 0.1 g and a resonance frequency of 10 kHz, we find $\delta x = 0.2$ nm. The corresponding change in capacity is $20 \cdot 10^{-18}$F $= 20$ aF (attofarad). Can we electronically detect such a small change in capacity? Certainly not when we use hybrid integration with bond pads, bond wires and an external ASIC.

A bond pad has a capacity of some pF, a million times more. Therefore, monolithic integration is a must.

Still, the task is extremely challenging. The trick is synchronous demodulation, as described in Chapter 1.3. An electronic circuit[3] for a capacitive accelerometer is shown in Figure 4.10. First, read the figure without the red dashed parts. The sensor itself is a differential capacity, which is driven by an HF oscillator at ω_{HF} in a half-bridge configuration. The frequency ω_{HF} is in the range of 1 MHz, far above the measurement bandwidth. The bridge output is an HF signal with amplitude proportional to δx. This is amplified and multiplied with the HF oscillation. This multiplication makes a synchronous demodulation: Signals at ω_0 are rectified; noise at any other frequency remains a HF signal, which is removed by the following lowpass filter.

In this configuration, we let the mass move and measure its amplitude. An alternative method is keeping the mass at its place and measuring the force we need. This can be done by adding a force balance loop with a negative feedback, as shown in Figure 4.10 in the red dashed part. When the mass wants to move up, a voltage is put on the mass, which generates an electrostatic force pulling the mass down. As a result, the mass does not move at all. How can we measure acceleration then? The voltage applied by the system to keep the mass in place gives the electrostatic force needed, and this must equal Newton's force. Thus, x_{out} denotes the acceleration. Actively controlled systems are more complicated, but there are important advantages:

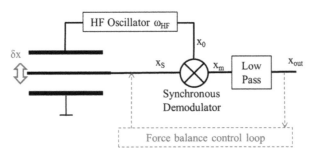

Figure 4.10 Electronics of a capacitive accelerometer. Synchronous demodulation is used to reduce the noise. Some systems additionally use a force balanced loop (red dashed line) to prevent movement of the mass by active control.

---

[3]S. D. Senturia: Microsystems Design, Springer ed. (2001).

- Like every spring, also the spring of the accelerometer will show non-linearity at higher movement amplitude. This nonlinearity is removed by active control since amplitude remains zero at all times.
- Individual variations in technology become less important for some technology parameters, e.g. the spring constant is not so important any more since the spring is not elongated anyway.
- The sensor can be faster. When we measure amplitude, after a step in signal, we have to wait until the spring mass system finds a new place, as shown in Figure 4.5. The active control starts as soon as the mass wants to start moving. Regulating back to zero is much faster than waiting for a new equilibrium.

In fact, for accelerometers, both methods are applied: amplitude measurement and force balanced control.

4.2.5 Resolution and Noise

How can we estimate the noise of an accelerometer? Nyquist resistor noise cannot be applied, there is no resistor. But the basic idea of noise due to thermal movement still holds. Like every mass, the seismic mass of an accelerometer will show thermal movement, and we can attribute an energy of $\frac{1}{2} k_B T$ to it. We now follow exactly the line of thought of Chapter 1.2.3., but we use mechanic terms instead of electric terms. Thermal movement of the mass will generate a noise velocity v_N. From this, we have to proceed to a power. Looking at the second term of the equation of resonance (4.3), we find a force proportional to the velocity and the damping b of the resonator.

$$F_N = b v_N \qquad (4.9)$$

Force and velocity result in power[4]

$$P_N = \frac{1}{4} F_N v_N = \frac{F_N^2}{4b} \qquad (4.10)$$

From Chapter 1, we know that the power is

$$P_N = k_B T \Delta f. \qquad \text{And thus} \qquad (4.11)$$

$$F_N^2 = 4 k_B b T \Delta f \qquad (4.12)$$

---

[4]T. B. Gabrielson: Mechanical-thermal noise in micromachined acoustic and vibration sensors. IEEE Transactions on electron devices (1993).

This describes the force exerted by thermal movement on the proof mass. Using Newton's formula, we proceed from force to acceleration and find

$$a_N^2 = 4k_B T \frac{b}{m^2} \Delta f = 4k_B T \frac{\omega_0}{mQ} \Delta f \qquad (4.13)$$

where Q is the quality of the resonator as defined in Equation 4.7. At this point, the mass comes back into play. As we have learned, in the case of a specified resonance frequency, large mass does not increase sensitivity. But it does increase the signal-to-noise ratio! A very small mass will show larger thermal movement and increase the noise. Hence, we find the mass in the denominator. Extremely sensitive accelerometers, such as geophones or microgravity sensors for spacecrafts, have large proof masses.

Assuming a mass of m = 1 μg and a quality factor of 0.5 we estimate the noise equivalent acceleration to be

$$\frac{a_N}{\sqrt{\Delta f}} = 60 \frac{\mu g}{\sqrt{Hz}} \qquad (4.14)$$

The data sheet[5] gives a specification of

$$\frac{a_N}{\sqrt{\Delta f}} = 80 \frac{\mu g}{\sqrt{Hz}} \qquad (4.15)$$

We can conclude that the simple thermal noise model describes the behavior of the system quite well.

Table 4.2 compares the argument for the thermal and mechanic cases. The line of the argument is parallel in both cases. There is a general rule about this, called the theorem of fluctuation and dissipation. The fluctuation is the quantity of measured noise (IR power, acceleration). This is correlated to the energy dissipation in the system (resistivity, friction) with a Nyquist-like formula.

The technology of accelerometers has made considerable progress in the last years. This can also be seen looking at the decrease of power consumption, as shown in Table 4.3. Power consumption is crucial for security and monitoring applications. Imagine you monitor the transport of sensitive goods, e.g. expensive optic devices. Temperature and humidity are measured once a minute; 99.9% of the time, the sensors are in sleep mode. But the accelerometer can never sleep. When there is a deadly mechanic shock, it is over in a millisecond and you cannot risk missing it.

---

[5]http://www.analog.com/media/en/technical-documentation/data-sheets/ADXL356-357.pdf.

Table 4.2 The noise of thermopiles and accelerometers and the theorem of fluctuation and dissipation

	Electric	Mechanic
Identify the specific way of thermal movement	Current	Velocity
Find the law of cause and effect	Ohms law: $U = RI$	Force of friction: $F = vb$ b = coefficient of friction v = velocity
Write down power	$P = \dfrac{U^2}{4R}$	$P = \dfrac{F^2}{4b}$
Thermal energy and power	$P_N = k_B T \Delta f.$	
Nyquist-type relation	$U_N^2 = 4k_B T R \Delta f$	$F_N^2 = 4k_B b T \Delta f$
Fluctuation = noise equivalent observed quantity	Voltage	Force
Dissipation	Electric resistance R	Mechanical friction b

Table 4.3 Development of ultra-low-power accelerometers

Year	Power	Bandwidth	Type	Resolution
2006	300 μW	10 Hz	ACB302	2 mg at 100 Hz
2010	90 μW	15 Hz	ADXL 345	
2013	3.6 μW	100 Hz	ADXL362	175 μg/$\sqrt{}$Hz
2017	3.4 μW		LIS3DH[6]	220 μg/$\sqrt{}$Hz
2018	2.6 μW	25 Hz	MC3610[7]	220 μg/$\sqrt{}$Hz
2018	21 μW	1.5–1000 Hz	BMA423[8]	140 μg/$\sqrt{}$Hz

4.3 Angular Rate Sensors

Angular rate sensors, also called yaw rate sensor or micro gyroscopes, measure the rate of rotation using an inertial force, the Coriolis force.

The dimension of angular rate is 1°/s. The earth rotates with 360°/day, which is 0.004°/s. This can be resolved only by the very best micro gyros. Fiberoptic gyros and mechanical gyros using spinning masses can resolve below 0.001°/s; they are used for navigation, but are very expensive. The realm of MEMS gyros is the regime between 0.005°/s and 0.1°/s.

---

[6]http://www.st.com/en/mems-and-sensors/lis3dh.html.
[7]http://www.st.com/en/mems-and-sensors/lis3dh.html.
[8]https://www.bosch-sensortec.com/bst/products/all_products/bma423.

4.3.1 Applications and Ranges

- Again, car industry is a driving market. A major security feature is electronic stability control (ESC), shown in Figure 4.11. Imagine you steer around a curve on an icy road. Suddenly the car slips, you steer, but there is no effect. The car leaves the course and you end up in the ditch or in the oncoming traffic. Can this disastrous situation be detected and counteracted? To detect, we compare the intended angular rate and the real angular rate. The intended angular rate is calculated from the speed and the steering wheel angle. The real angular rate is given by a micro-gyroscope. When they do not match, the system takes action on the brakes. When the car drifts toward the left, the brakes of right-hand side are actuated to bring it back on track. A gyroscope for ESC typically must resolve 0.1°/s.

  ESC and ABS systems are close relatives: The movement intended by the driver is analyzed and compared to real movement measured by inertial sensors. When they differ, the brakes are activated accordingly. While ABS (anti-blocking system) looks for linear acceleration, ESC looks for rotation.

- Advanced airbag systems use gyroscopes to detect turnover situations. When a car hits frontally against a wall, the accelerometer sees it and the front airbag is actuated. When a car hits an obstacle only with the fender,

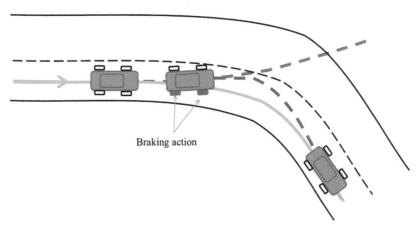

Braking action

Figure 4.11 The electronic stability control. When the car leaves the intended path, the brakes are actuated to bring it back on track.

the damaged car will rotate. The worst thing that can happen is a full turnover, often with fatal injuries. To prevent these, new airbag systems add side airbags and head airbags. To decide which airbag must be fired, gyroscopes are used for the detection of spinning and turnover situations. Here, the specs are also in the range of 0.1°/s.

- A third car application is the interpolation of GPS signals. In a tunnel or between large houses, the GPS system suffers from bad reception of satellite signals. This can be helped by interpolation: the location is calculated on the basis of known speed and direction. The next step is inertial navigation using a gyro. Then, even a curve in a tunnel can be correctly interpolated. GPS interpolation needs good gyros: 0.01°/s resolution are required.
- Drones and quadcopters need gyros for flight stabilization.
- Like accelerometers, gyros are used for man–machine interaction, such as game stations.
- Cameras and camcoders use gyros for image stabilization. When you zoom a camera on a faraway object and your hand trembles only slightly, the picture is blurred. This is prevented by optical image stabilization (OIS): the angular tremor is measured with a gyro. In the zoom optics, one particular lens can be moved by small actuators in x and y directions. The actuators exactly counteract your tremor, so that the picture on the CCD array is not shaky any more. Figure 4.12 shows that the power of those systems is impressive.

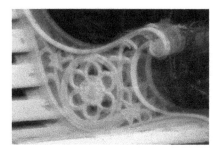 

The optical image stabilization (OIS) is off. You can clearly see how my hand was shaking.

When I switch on the OIS, I get a sharp image. The shaking is completely compensated.

Figure 4.12 Details of a garden bench taken at low light from 10 m distance using a strong tele lens.

4.3.2 Sensor Idea and Coriolis Force

When a reference system rotates, there will be inertial forces: the centrifugal force and the Coriolis force. Coriolis force is observed when, within a reference frame rotating with a rate Ω, a mass is moving radially inside out with a linear speed v_{rad}, as shown in Figure 4.13.

When the mass is moving force free, it will make a linear path. But for the observer on the rotating disk, this linear movement looks not linear since the reference frame is rotating. The apparent course is shown in red. The observer concludes that the mass is deflected from the linear path by a force acting to the side, which is called Coriolis force.

The deflection y is given by the tangential velocity $v_{tan} = \Omega r$ and the time t:

$$y = v_{tan}t = \Omega rt \tag{4.16}$$

The radius at time t is $r = v_{rad}t$, therefore

$$y = v_{rad}\Omega t^2. \tag{4.17}$$

The rotating observer explains this with the Coriolis acceleration a_C acting on the mass:

$$y = v_{rad}\Omega t^2 \rightarrow \dot{y} = 2v_{rad}\Omega t \rightarrow \ddot{y} = 2v_{rad}\Omega = a_C \tag{4.18}$$

The Coriolis force F_C is the acceleration multiplied by the mass:

$$F_C = ma_C = 2mv_{rad}\Omega \tag{4.19}$$

To allow for movements in all directions, we have to write a vector equation

$$\overrightarrow{F_C} = 2m\overrightarrow{v} \times \overrightarrow{\Omega} \tag{4.20}$$

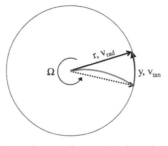

Figure 4.13   Within a rotating reference frame (rotating plate), a point moves inside out. Seen from outside, the point moves straight on. But, from the point of view of the rotating system, it is accelerated to the side. This is called Coriolis acceleration.

Note that this is a vector product: When the linear movement (v) is perpendicular to the vector of rotation (Ω), the Coriolis force acts to the third direction (tangential). For axial movement, the linear movement is parallel to the vector of rotation and the Coriolis force vanishes.

4.3.3 Layout

How can we make a micro sensor for angular rate? We need three movements in three perpendicular axes:

1. The external rotation Ω we want to measure.
2. The linear movement v (v_{rad}). We can realize this with a spring mass resonator, but we have to drive this motion actively. This motion is called the primary mode.
3. Coriolis force is going to drive a secondary mode, which is perpendicular to 1 and 2.

As for the accelerometer, we realize all components in one structure layer: spring, mass and displacement sensor plus actuator for the primary motion. The geometry is sketched in Figure 4.14. The mass is held by four springs, which allow x and y movement. Four comb drive actuators drive the primary motion in the x direction. The external rotation we want to measure is perpendicular to the plane of the paper. Then, Coriolis force will generate a secondary movement in the y direction. This movement is picked up with electrostatic electrodes.

Now we will analyze the double function of comb drives as actuators and as sensors.

First, we discuss an electrostatic actuator at the instance of a normal plate capacity (Figure 4.15).

When a capacity C is loaded to a voltage U, the stored charge Q is

$$Q = CU \tag{4.21}$$

And the stored energy E is

$$E = \frac{1}{2}CU^2 = \frac{1}{2}\frac{Q^2}{C}. \tag{4.22}$$

Now we imagine a small movement of the upper plate. There is movement and force, so a mechanical work is done, and there is a corresponding change in stored electric energy. Conservation of energy tells us that

$$F\,dx + dE = 0 \tag{4.23}$$

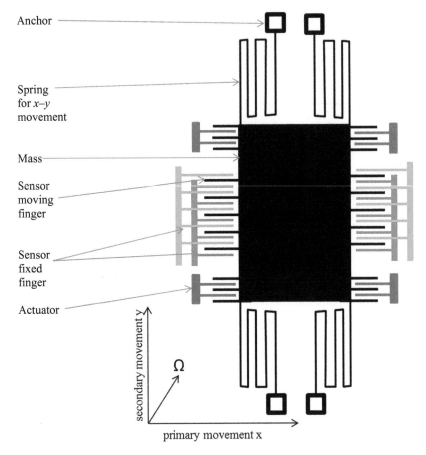

Figure 4.14 Schematic of an angular rate sensor.

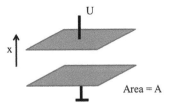

Figure 4.15 Force on the plate of a capacitor. The upper plate is movable, so increasing *x* means decreasing capacity.

Note that the capacity is disconnected, so the charge is constant, but capacity and voltage are not constant. The force is

$$F = -\frac{dE}{dx} = -\frac{1}{2}Q^2 \frac{d}{dx}\left(\frac{1}{C}\right) = \frac{1}{2}\frac{Q^2}{C^2}\frac{dC}{dx} = \frac{1}{2}U^2\frac{dC}{dx} \qquad (4.24)$$

For the plate capacity

$$C = \frac{\varepsilon_0 A}{x}; \quad F = -\frac{1}{2}\varepsilon_0 A \frac{U^2}{x^2} \qquad (4.25)$$

This type of actuator is strongly nonlinear since the force is going to the inverse square of the distance. Plate actuators are typically used in microfluidic valves or pumps. There, the initial distance is some micrometers. The actuator can handle large force, but only in a very short path.

Now let us consider a comb drive actuator as seen in Figure 4.16-A. The capacity is

$$C = \frac{h}{d}n\varepsilon_0(x_0 - x)$$

d: distance between the fingers
h: height of the fingers = thickness of the device layer
n: number of gaps

$$(4.26)$$

where $(x_0 - x)$ is the overlapping length of the combs. Thus, $h(x_0 - x)$ is the area of the capacity. The force in up-down (y) is cancelled, but there is a net force in-out, which is the x direction. When we calculate the force direction of x, we obtain:

$$F_x = \frac{1}{2}U^2 \frac{dC}{dx} = -\frac{1}{2}\varepsilon_0 \frac{nh}{d}U^2 \qquad (4.27)$$

The force is independent of x, which means that the actuator can exhibit a constant force while moving in and out. We can also realize a longer path, which is important for the primary movement of gyros. For this reason, comb drives are superior when we want to move small structures on a silicon chip. To increase force, we have to use a thick structure layer and a small gap between the combs. This is a technological challenge; it forces us to develop DRIE etching for large layer thickness h, which allows simultaneously a small gap d. The ratio of thickness to lateral precision is called aspect ratio.

When we use comb electrodes as <u>displacement sensors</u> measuring y, the vertical displacement in Figure 4.16-B, one moving finger makes two capacities with two fixed fingers (differential capacity). When the mass moves, at one side, the capacity rises, and at the other side it decreases. This can be realized using two fixed fingers on different potentials for each moving finger. Writing l for the overlapping length of the fingers, the capacity change is

$$C = n\varepsilon_0 \frac{lh}{\delta y} \qquad (4.28)$$

$$\frac{dC}{dy} = -n\varepsilon_0 \frac{lh}{\delta y^2} \qquad (4.29)$$

Also here, the aspect ratio of etching is the clue to good performance.

For movement in and out (*x*-direction), there is also a change in capacity, of course. We do not want to measure this change with *x*, because it corresponds to the primary movement, but we want to measure the secondary movement. What we do is to apply two symmetric sets of combs, as shown in Figure 4.14. Imagine a large primary motion in *x*, superimposed by a small secondary movement in *y*. The primary movement will reduce capacity in the combs on the left-hand side, but increase capacity in the combs on the right-hand side by the same amount. When we add the capacities, the net change vanishes. The secondary movement upward in *y* will increase capacities toward the upper (green) counter electrode on both sides and reduce capacities toward the lower (blue) counter electrodes. When we add, we will double the effect. In this way, the symmetric arrangement allows us to suppress primary mode response while we increase secondary mode response.

This type of displacement sensor is generally used in accelerometers. It allows small sensors, but it needs two layers: structure layer and a buried metal to allow the crossing of two lines (see Figure 4.16-B). We can avoid the buried metal with an alternative layout shown in Figure 4.16-C. Here, we separate the differential capacity, using one set of moving plus fixed fingers, which approach with rising *y*, and a second one for the other side. We need more space for the layout, but we can realize everything in one layer only.

For a real micro gyro, as shown in Figure 4.17, the design becomes much more complex. The first we see are the holes in the mass. These are etching holes for underetching the mass when the sacrificial layer is removed. But also the geometry is much more elaborate than in the simple sketch of Figure 4.14.

To understand the layout, we first have to discuss the idea of decoupling primary and secondary motion. The primary motion amplitude of such a gyro is a few micrometers, the secondary motion amplitude is <1 nm. This causes a major problem in actuation: Imagine that, due to some small geometrical imperfection in etching, the actuator is not perfectly in the *x*-direction, but slightly tilted with a small *y*-component. Then, there will be a small force component in the *y*-direction caused by the actuator. The system cannot distinguish this from a Coriolis force caused by rotation. Even an angular misalignment of 1/1000 degree would be enough to destroy the measurement. Effort needs to be taken to keep the primary actuator in line. This can be done by using not a 2D spring, but a 1D spring, which allows only the primary motion *x*. But we must not inhibit the secondary motion caused by the Coriolis force! This can be done using the idea of a decoupled mass. The actuator drives a U-shaped sledge, which can only move in primary

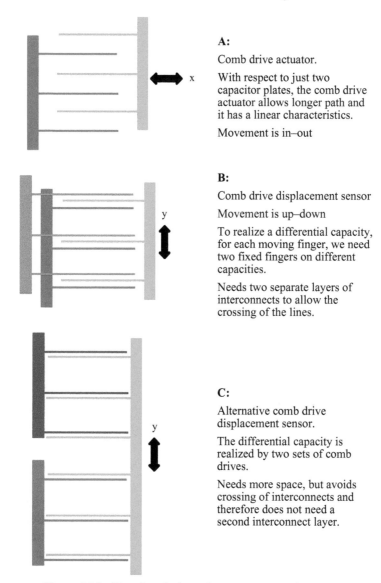

A:

Comb drive actuator.

With respect to just two capacitor plates, the comb drive actuator allows longer path and it has a linear characteristics.

Movement is in–out

B:

Comb drive displacement sensor

Movement is up–down

To realize a differential capacity, for each moving finger, we need two fixed fingers on different capacities.

Needs two separate layers of interconnects to allow the crossing of the lines.

C:

Alternative comb drive displacement sensor.

The differential capacity is realized by two sets of comb drives.

Needs more space, but avoids crossing of interconnects and therefore does not need a second interconnect layer.

Figure 4.16 Use of comb electrodes as actuators and as sensors.

motion x. Within this sledge, the real seismic mass is fixed by the secondary springs, which allow secondary movement in the *y*-direction. In this way, any *y*-oriented force component of the actuator is inhibited, yet the seismic mass is free to move in the *x* and *y* directions.

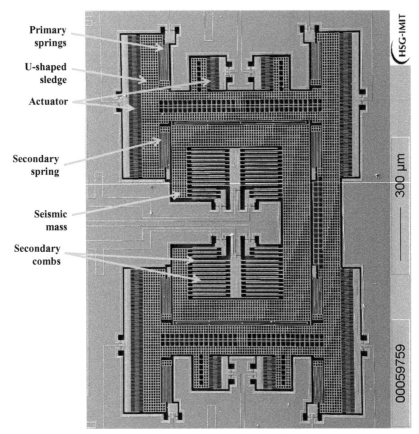

Figure 4.17 A picture of a micro gyro. Picture by courtesy of M. Trächtler, HSG-IMIT, Villingen-Schwenningen. The device layer from monocrystalline silicon is 30 μm thick.

Figure 4.18 shows the primary spring in detail. The spring is free to move in the x direction (horizontally), but it is hard with respect to y, the direction of secondary movement. Why the U-shape? This is done to prevent temperature problems. Consider two resonators as shown in Figure 4.19, a linear one and a U-shaped one. Imagine a rise of temperature, causing different thermal expansion of the substrate and the structure layer. In the linear resonator, the linear spring will be stressed and the resonance frequency will change like for a detuned guitar string. For the U-shaped resonator, a small displacement of the anchor will have only a negligible impact on movement and the resonance frequency will not change.

When the linear spring is deflected, the restoring force has two components:

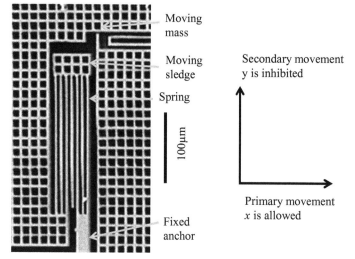

Figure 4.18 Details of the primary spring.

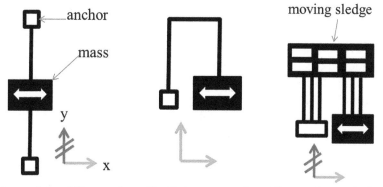

Two anchors, stiffness and elongation mode. Movement in *y* is forbidden, but due to the elongation mode, we will have temperature problems ("guitar string effect")

The U-shape uses only the stiffness mode, no temperature problem, but movement in x and y allowed.

Three springs each side and a moving sledge. Moving in *y* is forbidden. This type of spring is used as primary motion spring in the gyro.

Figure 4.19 How to fix a mass that is meant to allow movement in *x*, but not in *y*.

- stiffness mode: like a bar, the material resists bending.
- elongation mode: like a guitar string, the material resists elongation.
  The elongation mode is subject to temperature problems. Hence, guitars and violins have to be retuned all the time. The U-shape layout takes its restoring force only from stiffness. There is no elongation mode, and resonance frequency is temperature-stable.

4.3.4 Technology

Coriolis force goes with the inertial mass, so we have to realize a certain mass to achieve a good gyro. Using surface micromachining means that a thick structure layer is needed. First micro gyros[9] in the 1990s made using thick polysilicon as for accelerometers could achieve a resolution of 0.1°/s. For better resolution, new methods with thicker process layers had to be developed. The solution was silicon-on-insulator (SOI) technology. One silicon wafer is bonded on a second one with a thin oxide layer in between. The upper wafer is thinned down to the desired thickness. This SOI substrate is the starting point of the gyro process shown in Figure 4.20.

A micromachining process similar to the one given in Figure 4.20 is offered by X-FAB under the name XM-SC[10]. The process is carried out on 6-inch wafers, the device layer thickness is 15 μm and the trench width and beam width are both 2 μm.

For accelerometers and gyros, three big technological challenges had to be mastered. In this way, inertial sensors have been a major driving force for micromachining technology development from 1990 onward.

1. The sticking problem
   When two surfaces touch, there is always a mutual attraction due to electrostatic dipole interaction, which is called van der Waals force. When a small cantilever touches ground, it may stick there, like a little piece of plastic foil sticks on your fingers due to electrostatics. Once this happens, you cannot remove the cantilever any more, and the inertial sensor is lethally blocked. A number of methods have been published to avoid sticking, such as special etching procedures or monomolecular films.[11,12] Today, sticking can be prevented, but it seems to be a major company secret everywhere. The producers of inertial sensors normally do not publish what method they actually apply.

---

[9]New designs of micromachined vibrating rate gyroscope with decoupled oscillation modes. W. Geiger, B. Folkmer, U. Sobe, H. Sandmaier, W. Lang. Sensors & Actuators A 66 118–124 (1998).

[10]https://www.xfab.com/fileadmin/X-FAB/Download_Center/Technology/MEMS/Inertial_Sensor.pdf.

[11]N. Tas, T. Sonnenberg, H. Jansen, B. Lengtenberg, M. Elwenspoek: Stiction in Surface Micromachining. J. Micromech. Microeng. 6 (1996).

[12]W. Lang: Silicon microstructuring technology. Materials Science and Engineering Reports, Vol. R17 (1996).

2. DRIE etching with high aspect ratio

Using SOI technology, we can make a thick silicon device layer without problem, but we have to structure it. We want a high capacity of the interdigital structures, which means small gaps. At the same time, we

SOI-substrate (silicon on insulator), structure layer 15 μm; buried oxide, 0.6 μm. Doping for contact. Deposit and structure aluminum. In this process, we first make the bond pads. The reason is that, when the deep etching is done, we cannot use photoresist any more. We would not be able to get it out of the trenches.

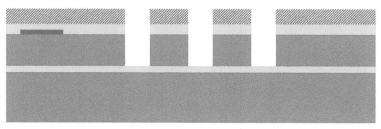

Deposit and structure PECVD oxide as a mask. Do not remove the resist yet. DRIE etch the wafer with the buried oxide as etch stop.

Remove the resist. Deposit PECVD oxide to passivate the side walls. Remove oxide at the bottom of the trenches by anisotropic RIE.

Figure 4.20   The process to make a micro gyro using SOI wafers.[13]

---

[13]W. Geiger, W. U. Butt, A. Gaißer, J. Frech, M. Braxmaier, T. Link, A. Kohne, P. Nommensen, H. Sandmaier, W. Lang: Decoupled Microgyros and the Design Principle Daved. Sensors and Actuators A 95 (2002).

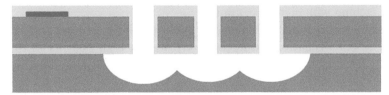

Change the etching modus to isotropic Si etching (chemical etch attack by reactive radicals). Etch deeply in the device wafer to underetch the structures to set them free to move. The anchor places are made broader in the layout so they are not completely underetched.

Remove the SiO$_2$ by isotropic plasma etching.

Prepare a wafer with caps by bulk micromachining in KOH. The wafer must have holes above the bond pads to allow electric contact. Deposit a thick layer of Pyrex glass on the cap wafer for anodic bonding.

Anodic bonding the cap wafer on the sensor wafer as a full wafer process under vacuum. Vacuum is important to maintain low damping of the resonators over a long time. Saw the whole sandwich. The capping protects the movable parts against saw dust and water.

Figure 4.20 *Continued.*

want to have a thick device layer to increase mass. The ratio of trench depth h to trench width d is the aspect ratio AR:

$$AR = \frac{h}{d}. \tag{4.30}$$

Wet chemical etching allows an aspect ratio of AR~1. First DRIE processes achieved an aspect ratio of 4. In gyro production, today aspect ratios from 6 to 10 are commonly applied, and the XM-SC process by X-Fab uses 7.5 (15 μm thickness, 2 μm trenches). In research, modern etching equipment can reach much higher values. Today's DRIE etching machines can etch a 3-μm-wide trench with an aspect ratio of 97.[14] The increase of aspect ratio in DRIE etching is the main reason for the improvement of gyro resolution in the last 20 years.

3. Full wafer encapsulation
   The usual way to saw a wafer in chips is to fix it on an adhesive foil, then saw the silicon with a diamond blade, and then pick the chips from the foil. For inertial sensors, this will not work. Once a corn of sawdust penetrates between the moving combs, then the sensor is blocked. After many experiments, the solution was full wafer bonding, as shown in Figure 4.20. We close the structures in the clean room on wafer level. Then, we can saw and clean without any harm. Full wafer level packaging also solves another problem: the long-term enclosure of vacuum. Gyroscopes need to work in vacuum, otherwise the damping will inhibit the primary motion. Long-term stable evacuation can be done by anodic bonding of the cap wafer under vacuum conditions.

4.3.5 Movement and Noise

To analyze the motion of a gyro, we write the Coriolis force. The primary motion is denoted by x, and the secondary motion by y.

Primary movement is $x = x_0 sin\omega_0 t$. Assuming a driving voltage of 2 V, a resonance frequency of 10 kHz and a quality factor of 1000, we can realize a primary amplitude of $x_0 = 5\mu m$. This shows us that in the case of the gyro, we must have a high quality factor, i.e. a low damping. Critical damping would stall the primary motion.

The secondary motion y is driven by the Coriolis acceleration

$$a_C = 2\dot{x}\Omega = 2\Omega\omega_0 x_0 cos\omega t = a_{C0} cos\omega t \tag{4.31}$$

---

[14]K. J. Owen, B. van der Elzen, R. L. Peterson, K. Najafi: High aspect ratio deep silicon etching. IEEE MEMS Conference (2012).

with

$$a_{C0} = 2\Omega\omega_0 x_0 \qquad (4.32)$$

For an external rotation of 0.1°/s, the secondary amplitude is $y_0 = 0,01$ nm.

This is far less than the distance of two atoms in the silicon material! Primary and secondary amplitudes have a ratio 0.01 nm to 5 μm, which is 1:500.000. This shows why decoupling is necessary.

The thermal noise of the gyroscope is easy to calculate since we already know the thermal noise of a resonating mass from the case of the accelerometer. To find the noise equivalent angular rate Ω_N, we put the noise acceleration from Equation 4.13 equal to the Coriolis acceleration:

$$a_N^2 = 4k_B T \frac{\omega_0}{mQ}\Delta f = a_C^2 = 4\Omega_N^2 \omega_0^2 x_0^2 \qquad (4.33)$$

$$\Omega_N^2 = \frac{k_B T \Delta f}{\omega_0 x_0^2 mQ} \qquad (4.34)$$

To get a good gyro, we need a high resonator quality Q (=small damping), large mass, large primary motion and large primary frequency. For a realized device W. Geiger[15] gives Ω_N = 0.04°/s at a bandwidth of 50 Hz.

How does this noise compare to experiment? To measure the noise, the output signal of the gyro is electronically filtered (Δf = 50 Hz), a number of N = 15,000 values x_i are measured. The 1σ noise equivalent angular rate is calculated as the standard deviation of the measured x_i by

$$\Omega_N = \frac{1}{R}\sqrt{\frac{1}{N-1}\sum_{i=1}^{N}(x_i - \bar{x})^2} \qquad (4.35)$$

where R is the sensitivity of the device. In fact, the device has an RMS noise of 0.025°/s at 50 Hz, even somewhat better than our model calculation.

Yet, when we look closely and analyze the noise in the frequency range, we will find an increase of noise at low frequencies, which cannot be explained by our thermal noise approach in Equation 4.33. We have to look for other possible noise sources. Besides thermal noise, two other sources of noise have to be taken into account. Current noise (or Schottky noise) stems

---

[15]W. Geiger, W. U. Butt, A. Gaißer, J. Frech, M. Braxmaier, T. Link, A. Kohne, P. Nommensen, H. Sandmaier, W. Lang: Decoupled Microgyros and the Design Principle Daved. Sensors and Actuators A 95 (2002) 239–249.

from the fact that electric current is not continuous, but quantified in single elementary charges with the amount of $e_0 = 1.6 \times 10^{-19}$C each. For small charges and currents, this quantification results in a measurable noise. This can be visualized with the rice-pile model[16] of material flow: Imagine you drop rice grain by grain on the pan of a scale until the rice-pile flows over the edge of the pan. In average, for each grain falling on the pan, there will be one grain falling off the pan. Looking at the reading of the scale, you see the medium weight of the rice pile with a chaotic noise superimposed.

From time to time, there will be avalanches in the slope of the rice-pile. Then, many grains will fall off the pan simultaneously. These large but rare events generate <u>1/f noise</u>, which is observed in all measurement devices and electronic systems. It is correlated to critical phenomena, when a small impact may cause a large movement, like the set off of an avalanche. The noise spectrum of 1/f noise shown like in Figure 4.21, Figure 4.22 gives a practical example from a sensor system.

4.3.6 Inertial Measurement Units (IMU) and Sensor Fusion

For inertial navigation, we need six degrees of freedom: three axes of acceleration and three axes of rotation. By integration, the speed vector, the orientation and the location in space can be found. Small hybrid IMUs (inertial measurement unit), which integrate the sensors and the electronics, are commercially available.

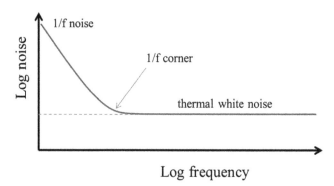

Log frequency

Figure 4.21 Typical noise spectrum of a measurement system. At low frequencies, 1/f noise is dominant, and at high frequencies, thermal (Nyquist) noise is dominant. The 1/f corner is typically at 10 Hz ... 100 Hz.

---

[16]Per Bak: How nature works. Copernicus ed. 1996.

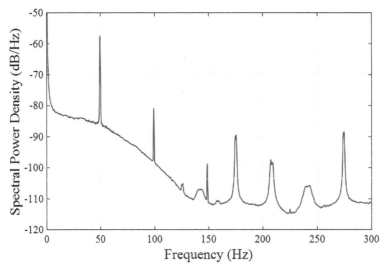

Figure 4.22   Measured noise spectrum of voltage measurement using an ADC converter[17]. The signals at 50, 100 and 150 Hz stem from the electric power grid. 1/f noise is clearly visible, the 1/f corner is at 140 Hz. Figure by Paul Meilahn.

In inertial measurement, a small bias in gyro baseline will add to a large deviation in path or angle after some time due to the integration. A bias of 0.01°/s is really small, but in worst case, it can add up to 180° after 5 h. This is the "true north" problem of inertial navigation: gyros are good in feeling curves, but they lose the true north position after some hours. To keep the true north, the systems are equipped with three-axis magnetic sensors as an electronic compass. By data fusion of the gyroscopes and the magnetometers, we gain the advantage of both sensor types:

- The gyro is good in detecting rotations, but it has no reference to the true north, and due to integrating baseline error, it will lose direction after some hours.
- The magnetometer sees the earth magnetic field and, in this way, has a long-term reference. On the other hand, it can be irritated by metal objects or electric currents.

Sensor fusion systems combine the two data in a way that on a short timescale, more weight is given on the gyroscope, while on a longer timescale, more weight is given on the magnetometer.

---

[17]Paul Meilahn, Bachelor thesis, IMSAS 2016.

An example of a nine-axis IMU is the FSM9 by Hillcrest labs[18]. The most important system data of this IMU are:

Size	19 × 18 × 4 mm	Resolution:	
Bandwidth sensor data	500 Hz	Gyro	0.04°/s
Bandwidth sensor fusion	250 Hz	Accelerometer	<6 mg
Power	2 V, 20 mA	Magnetometer	<1 μT

Sometimes there are two (or more) possibilities to measure one physical quantity, each having its weakness and strength. You can combine two sensor principles to get the advantages of both by sensor fusion.

A second example for sensor fusion is the Gyro-Inclinometer: To measure the true vertical in a car or an airplane doing a curve, an acceleration inclinometer is combined with a gyro:

- The inclinometer is referring to the vector of gravity, but it is irritated by the centrifugal force in the curve.
- The gyro is not sensitive to the centrifugal force. But it has no reference to the vector of gravity, and due to integrating baseline error, it cannot keep the true vertical for a long time.

Sensor fusion systems estimate a value (inclination, orientation, etc.) and correct the estimation continuously by comparing it to the input from two (or more) sensors working with different physical principles. In this way, they can combine the advantages of both measurement principles.

Questions

- How does an accelerometer work?
- Write down Newton's force equation.
- Explain the resonance curve of a spring-mass system. What is the quasistatic regime? What is critical damping?
- Draw the geometry of a silicon accelerometer (structure layer).
- Draw a comb drive displacement sensor.
- Estimate the noise of an accelerometer.
- Explain the Coriolis force. Write down Coriolis force equation and explain it.
- How does a micro gyro work?

---

[18]https://www.hillcrestlabs.com/downloads/fsm-9-data-sheet.

- How can you estimate the noise equivalent angular rate?
- What is an SOI wafer?
- What is 1/f noise?
- What is the difference of electrostatic actuators of plate type and comb type?
- Write down and explain the force on a the plate of a capacity

5

Magnetic Sensors

5.1 Dimensions, Ranges and Applications

The magnetic field B is defined by the Lorentz force on a moving charge, Figure 5.1:

$$\vec{F} = q\vec{v} \times \vec{B} \tag{5.1}$$

The dimension of the B-field is the Tesla $1\,T = 1\dfrac{Vs}{m^2} = 1\dfrac{N}{Am}$ or the Gauss: $1\,G = 10^{-4}$ T.

In this book, I will use the term magnetic field for the H-field and the B-field.

Both are correlated by the magnetic permeability of vacuum $\mu_0 = 1,26 \times 10^{-6}\frac{Vs}{Am}$ and the dimensionless relative permeability of a material μ.

$$B = \mu\mu_0 H \tag{5.2}$$

The dimension of the H-field is the A/m.

A strong permanent magnet can generate 1 T, for example, in the air gap of a loudspeaker. The earth magnetic field at the equator is about $0.5\,G = 50\,\mu T$.

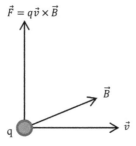

Figure 5.1 An electric charge moving in a magnetic field is subject of the Lorentz force. This is used to define magnetic field strength.

Position, movement, proximity, electric current and several other quantities can be measured indirectly by measuring the magnetic field. An example is the verification of the status of a door closed/open with a Hall probe or the rotation counting of a turning axis by putting a small magnet on the axis and using a Hall probe.

5.2 The Hall Sensor

When a current flows through a piece of material in a magnetic field, the Lorentz force is acting on the charge carriers. They will deviate from their straight way and concentrate at one side of the probe. This effect can be measured by applying electrodes at the probe and measuring the voltage. It is called the Hall effect, and the voltage is called Hall voltage. Figure 5.2 shows the geometry.

For this geometry, the current, field and voltage pickup are perpendicular, and hence, we only have to consider one vector component:

$$F_m = qvB. \tag{5.3}$$

When the electrons accumulate at one edge of the probe, the local accumulation of electrons on one side and the depletion on the other side will cause an electric field E, which generates an electrostatic force $F_e = qE$ on the electrons. The process is stationary when the magnetic and electrostatic

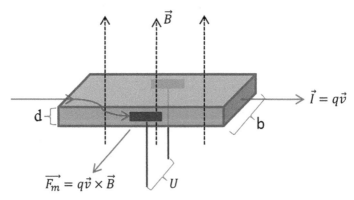

Figure 5.2 The Hall effect: b is the width and d is the thickness. A = bd is the cross section of the current flow. Moving charges are driven to the front by Lorentz force. This is measured by two electrodes.

forces cancel:

$$F_m = F_e. \tag{5.4}$$

Then, $vB = E$.

The width of the probe, or the distance of the electrodes, is denoted by b. The electric field is measured by the voltage

$$U = Eb = bvB. \tag{5.5}$$

The velocity of the electrons is given by the measurement current I we apply and by the number of movable electrons per volume n:

$$I = nvqbd. \tag{5.6}$$

Now, we can calculate the Hall voltage as

$$U = \frac{IB}{nqd} \quad \text{or} \quad U = A_H \frac{IB}{d} \tag{5.7}$$

where $A_H = \frac{1}{nq}$ is called the Hall constant.

What material should we use for a Hall probe, metal or semiconductor? A metal has many movable electrons, and thus large n and will generate a small Hall voltage. In a semiconductor, less electrons are movable, n is small and U becomes larger. If only a few electrons are moving, to transport a given current, they will have to move with a higher speed. This gives high Lorentz force and high Hall voltage. In the language of semiconductor physicists, the mobility of the electrons is higher in semiconducting material than in metal. Silicon has a high Hall constant, but there are other semiconductors that show higher electron mobility and therefore higher Hall constant. Typical materials for Hall sensors are indium antimonide (InSb) or indium arsenide (InAs).

It is worth having a second look on the sign of the Hall voltage. For the argument of charge carrier accumulation in Figure 5.2, we did not need to know the sign of the charge carriers. When the current is transported by negatively charged electrons, they will move from right to left. When the charge is transported by positively charges holes, they will move from left to right. The product of velocity and charge is the same, and the charges always accumulate in front. This means that, if the current is transported by electrons, the front electrode will become negative. If the charge is transported by holes in a semiconductor, then the front electrode will become positive. Hall constants for different materials can be positive or negative. Furthermore, measuring the Hall voltage tells us whether a given material is an electron conductor (negative Hall constant) or a hole conductor (positive Hall constant).

Hall sensors normally are monolithically integrated with electronics as Hall threshold switches or as analog sensors. The range is 0.1 mT to 10 T, and resolution down to 0.1 µT is possible. Some examples of Hall sensor applications are:

- In cars, we find up to 100 Hall sensors:
    - Has the safety belt properly been fastened?
    - Are all doors closed?
    - What is the position of the seats?
    - What is the rotation rate of each wheel?
    - What is the status of the gear lever?
    - How far in or out is each bumper (chassis position sensing)?
    - What is the angle of the steering wheel?

Most of this could be done by micro switches, of course, but switches are vulnerable to dust and humidity, while Hall sensors are reliable.

- Measuring electric current: Each current causes a magnetic field. This is measured by a Hall sensor. The advantage is that you must not open the circuit to mount the sensor. And the second advantage is that it works for AC and for DC, while current transformers only can measure AC. Battery control in cars is done using Hall sensors for current measurement.
- Analog multiplication: We can build an analog current multiplier when we use one current for the magnetic field and the other for the Hall sensor.
- Electronic compasses and navigation systems. Hall sensors with a resolution of 1 µT can be used for compasses.
- Whatever moves can be controlled by Hall switches. Examples are rotating gears or turbine blades. An important application is the control of brushless electric engines. Figure 5.3 shows an electric drive for a magnetic memory disk, taken from a PC. There is a base plate (left-hand side) with nine coils to generate a magnetic rotary field. The rotor (right hand, turned upside down) has a permanent magnetically coded ring. This ring follows the rotary field and rotates itself. To control the rotary field, we must instantaneously know the exact rotation angle of the magnetic coding of the ring. This can be measured by two Hall sensors on the base plate.

Figure 5.3 Electric drive of a magnetic memory disk. A magnetic ring is driven by the rotary field of nine coils. To measure the movement of the ring, two Hall sensors are mounted on the base plate. Their signals are used to control the rotary field.

- A really fascinating example is the control of a wheelchair.[1] People suffering from tetraplegia cannot do any movement with arms or legs. How can it be made possible for them to move on their own? One answer is the Tongue Drive System: A small permanent magnet is fixed in the tongue of the patient as a tongue piercing. Hall sensors are used to detect the movement of the tongue. The signals are used to control an electric wheelchair. A video of GT-Bionics Lab[2] shows how a tetraplegic patient moves around steering the wheelchair with his tongue.

5.3 The Resistive Hall Probe

The Hall effect can also be used to generate a magnetoresistive material, the Hall resistance probe. Figure 5.4 shows the structure of the material. In a semiconducting matrix with high Hall coefficient such as indium antimonide (InSb), there are thin metallic needles embedded. The current goes perpendicular to the needles. When the electrons enter the material, they are driven

---

[1] https://www.ncbi.nlm.nih.gov/pmc/articles/PMC4454612/
[2] GT-Bionics Lab, Georgia-Tech: http://gtbionics.ece.gatech.edu/

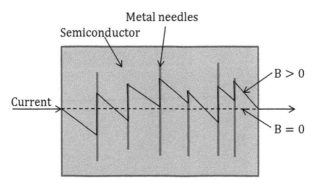

Figure 5.4 A resistive Hall probe with metal needles embedded in a semiconducting matrix. Moving particles are forced in a zigzag path; thus, the resistance they undergo is increased.

to one side by Lorentz force. When they enter a metallic needle, the Lorentz force is small and the electrons are driven back by the electric field generated by their accumulation at one side. This goes on and on and the electrons are going to walk on a zigzag line through the material. This increases the path length and thus the resistance with increasing magnetic field.

5.4 The Giant Magnetoresistive Effect

Small magnetic sensors are extremely important for data storage. In hard disk drives, information is stored in magnetic coding of the disc. To get more information on the area, extremely small and sensitive magnetic sensors are needed. This task has been solved by the Giant Magnetoresistance (GMR effect), an effect which gives a high change of electrical resistance when a magnetic field is applied.[3] GMR material is a multilayer of ferromagnetic and nonmagnetic thin films shown in Figure 5.5. When the nonmagnetic inter-layer is very thin, then the magnetic moments of the two ferromagnetic layers interact through this layer.

For a specific thickness of about 1 nm, the natural way the magnetic moments align without an external magnetic field is antiparallel. Yet, when a small external magnetic field is applied, this filed is going to bring the layers in parallel alignment. Why does this change electric resistivity? The point is that a travelling electron can be scattered in a magnetic film, and

---

[3]G. Binasch, P. Grünberg, F. Saurenbach, W. Zinn: Enhanced magnetoresistance in layered magnetic structures with antiferromagnetic interlayer exchange. Physical Review B, Vol. 29, No 7, March 1989.

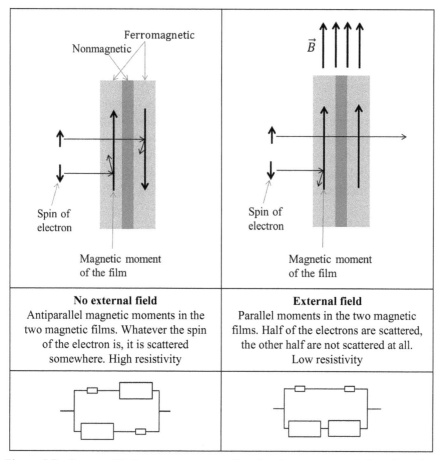

Figure 5.5 One possible layer structure to realize the giant magnetoresistive effect. One possible layer structure is Fe (ferromagnetic) and Cr (nonmagnetic). Bottom: two current model of conduction.

this scattering depends on spin. When the spin of the electron and the magnetization of the film are parallel, there will be little scattering. When they are antiparallel, there will be strong scattering. Now we can analyze the scattering of electrons travelling through the film for the two alignments, as shown in Figure 5.5.

Without magnetic field, the magnetic moments are antiparallel. When an electron is not scattered in the first layer, it will surely be scattered in the second. One of the layers always has the magnetic moment which allows scattering.

When a magnetic field is applied, the magnetic moments are aligned parallel. Then, half of the electrons are scattered everywhere. But the other half is lucky, i.e. it can pass through. Thus, the two alignments show a large difference in electric conductivity.

A helpful concept is the two-current model for current ferromagnetic metals. We consider the current split in two parallel currents, one with electron spin up and the other with spin down. Spin parallel to the magnetic moment of the film results in small resistance, and spin antiparallel results in large resistance. Then, we can draw the resistor networks shown in Figure 5.5 and see how parallel magnetization of the two ferromagnetic layers result in less resistance of the device.

5.5 The Tunneling Magnetoresistive Sensor

Magnetic orientation of layers can also influence the chance of an electron to tunnel through an insulating barrier. This is called tunneling magnetoresistance (TMR effect).[4] It gives a much stronger sensor response than the GMR.

What is tunneling of electrons? When an electron with an energy E hits on a potential barrier with height V>E, the electron cannot pass the barrier. Insulating layers are such barriers that do not allow electric current. But when the barrier is very thin, there is a chance that the electron can penetrate it, which is called tunneling. When a very thin barrier is placed between two magnetic layers, the tunneling chance is influenced by the relative orientation of the moments of these two layers. For a tunneling event, there are two preconditions: (1) there must be an electron that can tunnel and (2) there must be a state it can tunnel to. In the words of quantum mechanics: the transition probability is proportional to the product of the densities of states (DOS) of the emitting and receiving layers. In Figure 5.6, the DOS is drawn for electrons in the ferromagnetic films. If the spin is parallel to the film magnetization, then this density is high. If it is antiparallel, then the density is low.

There are four possible transitions numbered 1 to 4 in Figure 5.6:
Parallel alignment:

1. Down to down: Emitting and receiving states are sparsely populated: Very low current.
2. Up to up: Emitting and receiving states densely populated: <u>High current</u>.

Antiparallel alignment:

3. Down to down: Emitting states, but no receiving states: low current.
4. Up to up: Receiving states, but no emitting states: low current.

---

[4]http://www.dowaytech.com/en/1776.html

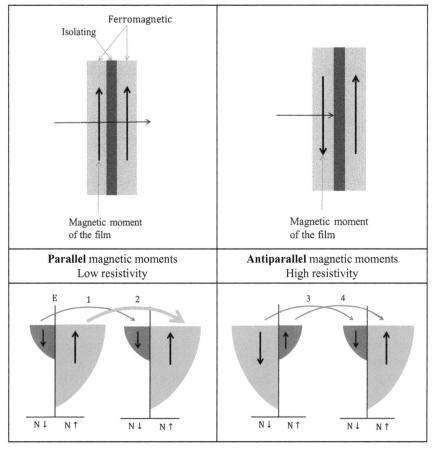

Figure 5.6 A tunneling magnetoresistive layer sequence. A possible material combination is iron (ferromagnetic)–manganese oxide (isolating tunneling barrier)–iron. In the lower part the Fermi level density of states N is drawn for the different electron spins.

Tunneling up to down and down to up will not happen since electrons do not change their spin while tunneling. So clearly the parallel alignment will allow tunneling current while the antiparallel will not.

Figures 5.5 and 5.6 just show simplified examples for one possible realization of GMR and TMR sensors. There are more layers, and there are many other possible geometries and material combinations.

Now we understand that mutual orientation of the two ferromagnetic layers will determine the tunneling current. But why should they be differently oriented? For a sensor, we have to fix one ferromagnetic layer somehow

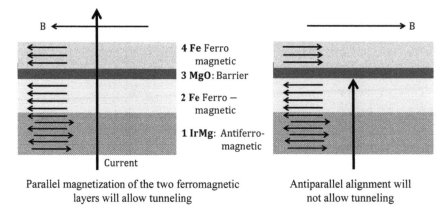

B ← ————————————   ————————→ B

4 Fe Ferro
      magnetic

3 MgO: Barrier

2 Fe Ferro −
      magnetic

1 IrMg: Antiferro-
      magnetic

Current

Parallel magnetization of the two ferromagnetic   Antiparallel alignment will
layers will allow tunneling   not allow tunneling

Figure 5.7   A tunneling magnetoresistive layer sequence. In real devices, there are more thin layers for magnetic shielding. For details, please refer to Zhu and Park.[5]

while we let the other one go with the external field. This brings us to the next concept: magnetic pinning. Figure 5.7 shows a layer sequence with one more layer at the bottom, which is antiferromagnetic. Antiferromagnetism means that there are domains with a certain magnetization, but these domains do not align parallel (as they do in ferromagnetism) but antiparallel. Here, the domains are horizontal sublayers. The purpose of this antiferromagnetic layer (layer 1 in Figure 5.7) is to define a stable magnetic direction. The first ferromagnetic layer (layer 2 in the figure) will align with respect to layer 1. This is called magnetic pinning.

Layers 2, 3 and 4 define the tunneling device as discussed in Figure 5.6.

Layer 2 is ferromagnetic, e.g. iron, and its moment is pinned.

Layer 3 is the tunneling barrier, e.g. manganese oxide. This layer stops the current unless tunneling becomes possible.

Layer 4 is ferromagnetic and not pinned. Its moment rotates according to the external field.

TMR sensors are increasingly replacing all other magnetoresistive material. The big advantage is the large effect: In GMR sensors, a magnetic field can change resistivity by 10%, and in TMR sensors, it can change by 100%.

The key application of the magnetoresistive sensors is magnetic data storage. The better the resolution of the magnetic sensor, the less area is needed for a single pixel, and the more data can be placed per area.

---

[5]Magnetic tunnel junctions. J. G. Zhu, C. Park. Materials today, vol. 9 (2006).

5.6 The Fluxgate Sensor

The fluxgate method uses the saturation of a magnetic core to measure the field with high accuracy and good resolution. Remember what happens when an iron core is subject to an external magnetic field H. In the iron, there are regions of locally aligned magnetization. They are called magnetic domains. In a magnetic domain, all the magnetic moments are aligned, but one domain is not aligned with respect to a neighboring domain. Now put the iron core in an electric coil and apply a current to generate a magnetic field H. This will align the domains in one direction, and thus the magnetization B of the material rises. At a certain point, all domains are aligned, and further magnetization is no longer feasible, which is called magnetic saturation. This is shown in Figure 5.8. For the analysis of the fluxgate principle, we will make simplified approximation as shown in Figure 5.8, right-hand side: negative saturation, then a linear regime and then positive saturation.

What happens when we superimpose a small external magnetic field H? The curve will slightly shift to the right, but this will not be recognized generally. To measure this small shift, we have to find an arrangement where all other effects cancel. This we do with a system shown in Figure 5.9, called fluxgate magnetometer. We take two identical iron cores. Then, we wind a coil around both of them, winding counter clockwise around the first core, and then clockwise back around the second one. What happens when we apply a current? Both cores are magnetized to the same degree, but their magnetization has an opposite sign. When we add the two magnetizations, they should cancel each other. The sum of magnetizations is measured using a third coil wound around both cores. It picks up the sum of the magnetizations of the two coils, and we call it the pickup coil.

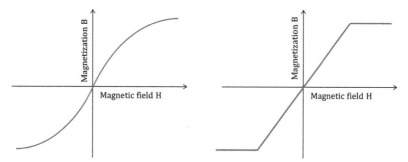

Figure 5.8 Magnetization curve of a weak ferromagnetic material showing the saturation & For understanding the fluxgate, we reduce the saturation curve to a straight line between the lower and upper saturations.

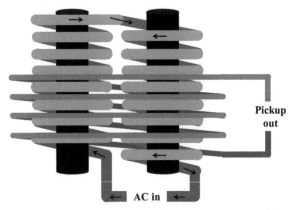

Figure 5.9 A fluxgate sensor. Two identical ferromagnetic cores with two windings of opposite sense of rotation. A third pickup coil measures the integral magnetic flux. With no external field, the magnetizations always cancel. With an external field, the magnetizations do not always cancel, which can be measured.
Figure adapted from J. A. Blackburn.[6]

Without an external field, the magnetizations of the two coils cancel exactly and the pickup coil will not pick up any signal. The two magnetization curves are shown in Figure 5.10 by the brown dashed lines. When we apply a current, one core goes from minus saturation to plus, while the other one goes from plus to minus. Applying AC, we drive the cores up and down, the absolute value of magnetization being always the same for both, but the sign opposite. Thus, the sum of the magnetizations is zero in any point of time, and the pickup coil will not give any signal.

Now imagine a small external magnetic field H superimposed, pointing upward. What will happen? As we discussed, the saturation curves will slightly shift to the right or left. For core 1, the external field helps us to drive, and the jump comes a little earlier. For core 2, the external field will counteract the driving current, and the jump will be a little delayed. These are the red solid lines in the figure. Now the two magnetizations do not cancel any longer.

To understand the pickup signal, we first look at the sum magnetic flux Φ. It is the sum of the two fluxes in each core. Following the curve from left to right, first we find zero since both cores are saturated and cancel. Then, core 1 starts to jump up, while core 2 still waits. The flux is going to rise.

---

[6]James A. Blackburn: Modern Instrumentation for Scientists and Engineers; Springer (2000).

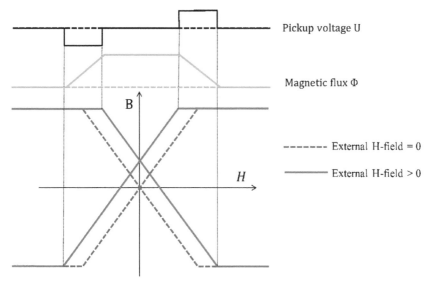

Figure 5.10 The magnetization curves in a fluxgate magnetometer. The X-axis is the H-field (external field plus field made by the excitation current). It is driven back and forth by the AC excitation current. Without an external field, the two cores have opposite B-fields (dashed lines). The sum is zero at all times, and no voltage is induced in the pickup coil.

When an external field is applied, one core moves up in B-field a little earlier since the external field acts in the same sense than the excitation. In the other coil, the external field acts opposite to the excitation and the moving up occurs a little later. The two fields do not cancel any more and a voltage is induced in the pickup coil.
Figure adapted from James A. Blackburn.

Then, core 2 starts to jump down. The flux is not rising anymore, but remains at a positive value. On the right-hand side, the reverse happens, and the flux goes back to zero.

The voltage induced in the pickup coil goes with the change of flux:

$$U = -n\frac{d\Phi}{dt}. \qquad (5.8)$$

The pickup voltage thus has two signals, one signal in plus, and a symmetric one in minus. Since the AC current drives magnetization back and forth, the x-axis in Figure 5.9 can be read as time axis and these two square signals will repeat for each AC cycle twice. This allows us to use a lock-in amplifier at double frequency to decrease the noise of the system.

Fluxgate magnetometers show a very good resolution in the μT range or lower. Bartington Instruments[7] make a magnetometer with a noise of $6\ pT/\sqrt{Hz}$ at 1 Hz. Fluxgates are used when you want to detect a small inhomogeneity in a magnetic field[8]:

- Medicine: Magnetic resonance imaging (MRI) for imaging the organs in the human body
- Geomorphology: For detecting different types of rock and sediments
- Geology: For analyzing the electric conductivity in the earth
- Forensic and archaeology: To detect hidden objects
- Electronic development: For measuring currents on circuit boards without interrupting the current path

Classic fluxgate sensors are made by winding copper wires. Of course, MEMS people tried to develop a planar fluxgate using thin film coils, and this actually works. First it is important to understand that the pickup coil must envelop the two iron cores, but not necessarily the two excitation coils. When an iron core is saturated, then the whole metal is saturated, also besides the excitation coils, and a layout as shown in Figure 5.11 will work. When we understand this, we can draw a planar design as shown in Figure 5.12,

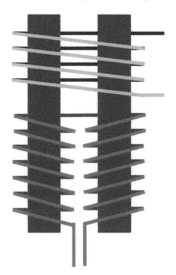

Figure 5.11   Alternative fluxgate structure.

---

[7]http://www.bartington.com/Literaturepdf/Datasheets/DS0061_Mag592_585.pdf
[8]http://www.bartington.com/market-sector-applications.html

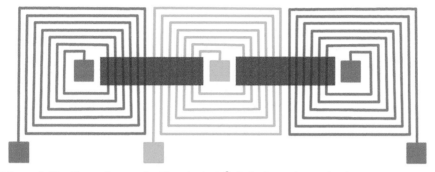

Figure 5.12 Planar fluxgate by T. Heimfarth.[9] Coils from electroplated copper, magnetic cores by electroplated NiFe.

which can be realized in thin film or in printed circuit technology.[10] There are three planar coils in a copper layer, and two magnetic cores in an iron layer on top of them. Without external field, the magnetization of the two cores will cancel. With external field, there will be a small phase difference as shown in Figure 5.10.

Questions

- Write down Lorentz force equation and explain it. Draw the vector tripod and explain it with the right-hand rule.
- How does a Hall sensor work?
- What material is better for Hall sensors: metals or semiconductors?
- How does a Hall resistive material work?
- Explain the GMR sensor
- Explain the TMR sensor
- What is magnetic pinning?
- Explain the fluxgate principle.
- Explain the magnetization curves in a fluxgate magnetometer and explain the induced voltage in the pickup coil.

---

[9]T. Heimfarth, M. Zubioli Meilli, M. N. Perez, M. Mulato: Miniature Planar Fluxgate Magnetic Sensors Using a Single Layer of Coils.

[10]M. Ortner, N. Navaei, M. Lenzhofer: Modelling Planar Fluxgate Structures. Eurosensors, Graz (2018).

6

Flow Sensors

6.1 Ranges and Applications

Flow is the movement of fluids, which can be liquid or gas. The flow velocity is a vector and has the dimension of meters per second (m/s). The local flow speed has mainly scientific interest. Practically, the flow rate through a tube is much more important. This is not a vector, but a scalar. Now we have to make an important distinction: do we want to measure the volume coming by, or the mass? The volume flow is measured in m^3/s, and the mass flow in kg/s. Volume and mass flow are correlated by the density ρ. Fluids are incompressible; for a given fluid, the density is a function of temperature, but not of pressure. Gases are compressible; the density depends on pressure and temperature. For technical application of gases, we often use the "standard liter per minute" (slpm) or the "standard cubic centimeter per minute" (sccm). Sccm gives the actual volume flow reduced numerically to standard conditions. Most times, the standard conditions are defined as $T = 0\,°C$ and $p = 1$ atmosphere $= 1.013$ bar.

Applications of gas flow rate measuring:

- Car industry: Measurement of the inflow of air to the combustion engine for the control of the amount of fuel to be injected to guarantee stoichiometric combustion. For combustion engines, it is very important to operate close to stoichiometric combustion. When there is too much air, the burning is too "lean". The combusted fuel reaches less temperature than possible, which reduces the thermodynamic efficiency of the engine. Furthermore, the amount of poisonous nitrogen oxide (NO_x) in the exhaust gas may rise, since there is not enough carbon around for full combustion of the oxygen. When there is too much fuel, the burning is too "fat". There are too many carbon atoms and not enough oxygen to form CO_2. Incomplete combustion results in the generation of carbon monoxide CO. To control stoichiometry, the engine has two sensor inputs: the amount of air intake is measured with a flow sensor, and the rest of oxygen in the exhaust gas is measured with a λ probe.

- Metering of gases for consumption billing and for process control.
- Aircraft speed: The speed of the plane versus the ground is measured with inertial sensing and GPS. Flow sensors are used to measure the speed of a plane versus the surrounding air. This is an important parameter for flight stability since the speed generates the uplift force of the wings. Insufficient flow around the wings can be very dangerous: not enough uplift, instable flight attitude. When there is strong tailwind this danger may be overlooked when looking only at inertial measurement. The pilot always controls the speed versus the air with a Prandtl tube and a pressure sensor. Iced or blocked airflow sensors have been causing several catastrophic plane crashes in the recent decades.
- Wind measurement: Not only meteorologists want to measure the wind, but also large buildings and industrial plants have wind sensors to identify dangerous storm situations. Wind energy plants measure the wind biaxial to put the rotor versus the wind and to control the pitch of the rotor blades.
- Air-conditioning can be optimized by flow and temperature information, allowing a considerable energy saving.

Liquid flow sensing is used for:

- Metering of water, hot water and fuel. Here, the challenge is to achieve a large dynamic range. The flowmeter might have a minimum detected flow of 50 l/h, and a maximum flow of 2500 l/h. This corresponds to a dynamic range of 1:50. Now consider the scenario of a tap leaking by 10 l/h. This goes unseen below the threshold, but it adds up to 87 m^3 per year, with a cost of some hundred Euros.
- Industrial process control: Media may be aggressive such as hot sulfuric acid. There we need sensors made of steel, silicon devices will not survive.
- Media can also be difficult in composition such as two phase flows: slurry, oil plus air, water plus sand, acrylic paint, mayonnaise, etc. In these cases all components build in the tubes such as heated wires or paddles will be destroyed.
- Biotechnology and analytical chemistry need flow sensors for very small flow rates. For a biomedical test, we may use a chip with a reaction chamber of size 10 μm. To control the filling of such a small volume with a flow sensor is very challenging.

6.2 Types of Flow Sensors

A number of macroscopic devices measure the flow by mechanic principles, such as small turbines, drag force moving up a probe body against gravity or filling of a defined volume (inverted pump). There are several macroscopic flow measurement systems, which are powerful but cannot be realized as micro sensors. One method is measuring the Doppler effect in ultrasound. Another method is to measure the speed of particles in the flow using a laser Doppler anemometer. We cannot cover all of these devices here. The systems we focus on are closely related to other sensors we have already been discussing: Venturi tubes are an interesting application of pressure sensors; Coriolis flowmeters allow us to repeat inertial forces; in electrodynamic flow sensing, we meet again the Lorentz force and the thermal flow sensor is almost a twin of the infrared thermopile. Thus, we learn about a fascinating chapter of sensors and, at the same time, we can repeat basic principles of this book.

6.3 Pressure Sensors for Flow Measurement

When a fluid moves, there is kinetic energy and there are forces involved, which generate a pressure difference. This can be used to measure the flow rate with a pressure sensor. The basic idea can be seen looking at the Venturi nozzle. This is a tube with a narrowing in its diameter, as shown in Figure 6.1. To understand the effect of flow on the pressure, we write Bernoulli's equation, which describes that the energy in a specific volume of the fluid is constant. Thus, the equation has the dimension of energy per volume, which is pressure ($J/m^3 = Nm/m^3 = N/m^2$).

$$p_0 + \frac{\rho}{2}v_0^2 = p_1 + \frac{\rho}{2}v_1^2 = p_{\text{tot}} = const \qquad (6.1)$$

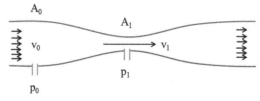

Figure 6.1 Venturi tube. When the diameter shrinks, the velocity increases. According to Bernoulli's equation, the pressure will decrease: $p_1 < p_0$.

The first term is the pressure energy. The second term is the kinetic energy. The sum of the energy is constant and is denoted as total pressure p_{tot}. The pressure p_1 is called the dynamic pressure.

Next, we have to look at the equation of continuity:

$$A_0 v_0 = A_1 v_1 \qquad (6.2)$$

The fluid is regarded as incompressible. Since the volume flow through the pipe is constant, a decreasing cross section A will cause an increase in velocity v. At the inlet, there will be a cross section A_0, a flow speed v_0 and a pressure p_0. When the cross section reduces to A_1, the flow speed will increase to v_1, and the pressure, according to Equation 6.1, will decrease to p_1. Finally, when the cross section opens to its initial area A_0 again, the speed reduces to v_0 again and the pressure increases to p_0 again. The last statement is only partially true since there will be friction in the narrow nozzle. Thus, some of the flow energy is transformed into heat and the outlet pressure will be a little lower. This we call the remaining pressure loss of the nozzle.

This is definitively against human intuition. When I squeeze the flow together, I would expect pressure rising. But intuition is not always true. Let us look at the energy: when the tube narrows, the flow must speed up. Kinetic energy goes up, and the energy must come somewhere. There is only one other energy which might be exploited: pressure. Therefore, pressure must drop when speed increases. There is a simple experiment to demonstrate this: take two sheets of paper in your hands and let them hang down parallel with a 5-cm distance. Then, blow between them. Will you blow them aside? No, quite contrary, they will come closer together, because the pressure between them reduces. Try!

One practical implementation of the Venturi tube principle is the orifice flowmeter shown in Figure 6.2. It is a straight tube with an orifice; this is a plate with a hole. How can we apply Bernoulli's equation here? Actually, the streamlines in this device will follow the geometry of a Venturi tube, as indicated in the figure. The streamlines will smoothly concentrate to a

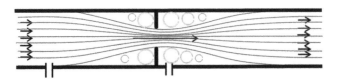

Figure 6.2 Orifice flowmeter. It has a simple geometry but it behaves like a Venturi tube since the flow lines run in the same way.

minimal point a little after the orifice and then open again. The rest of the volume is filled with vortices and acts as a deadwater area not influencing the stream lines. The advantage of the orifice configuration is that it is geometrically well defined and easy to fabricate. In the literature, you find exact geometries for the geometry of a standard orifice. Using this, you achieve very reproducible differential pressure values even without specific calibration of each sensor. This is a big advantage when an independent system for calibration measurement is not available.

Another implementation is the Prandtl tube (or Pitot tube) shown in Figure 6.3. It has two openings. The first is in front normal to the flow. Here, the flow lines are dividing to flow around the tube, which is called the stagnation point. There is vanishing flow velocity and therefore we can measure p_{tot} here. The second opening at the side is longitudinal to the flow, the air streams by and thus the reduction of pressure described by Bernoulli will happen here. From the differential pressure, we can calculate the flow speed v_0 at the second opening. Prandtl tubes are applied in every airplane, they are an important security feature to assure stable flight condition. You find them at the nose or under the wings of the plane.

Differential pressure flowmeters are the most common flow-sensing devices, but they have a serious drawback: the nonlinear characteristics of Bernoulli's equation (Figure 6.4). At low flow rate, the gradient is low. Thus, the sensor has small sensitivity for small flow rates. This causes problems concerning the dynamic range. Imagine you want to meter a maximum flow of 2500 l/h, and you want to resolve 25 l/h. This is a dynamic range of 1:100, not much for most measurement systems. But for the pressure sensor, things look different: when you use a 1 bar sensor, you have to resolve 0.1 mbar (1:10.000). This is hard to realize. On the other hand, for high flow rates, the gradient is steep. This causes the danger of breaking the membrane of the pressure sensor, when the flow becomes too fast.

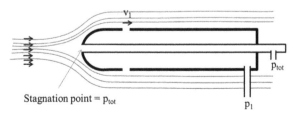

Figure 6.3 Prandtl probe to measure the flow around the probe, often used to measure the speed of an airplane with respect to the air. The head-on opening is a flow stagnation point, where we measure the total pressure p_{tot}. The openings at the side measure dynamic pressure p_1.

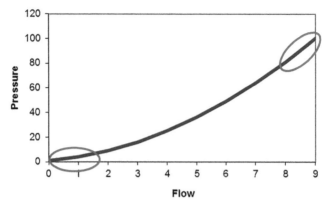

Figure 6.4   The characteristics of a Bernoulli-based flow sensor is quadratic. This caused two problems: At low flow, the sensitivity is small. At high flow, pressure increases fast and there is a danger of breaking the pressure sensor membrane.

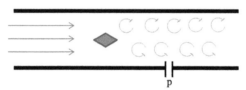

Figure 6.5   Vortex flowmeter. When the flow passes the obstacle, e.g. a wire, vortices are generated. The vortex frequency is measured with a pressure sensor, from which the flow speed is calculated.

Besides differential pressure measurement, a second method to use pressure sensors for flow measurement is <u>vortex shedding</u>, Figure 6.5. A flow around a small obstacle (e.g. a wire in a tube) will cause vortices. The frequency of the vortices is measured using a pressure sensor. It is a function of the flow speed.

6.4 Coriolis Flow Sensor

In a Coriolis flowmeter, a U-shaped tube is made to vibrate up and down by actuators, as shown in Figure 6.6. In practical systems, this primary vibration will have an amplitude of about 10 μm. The vibratory movement is measured for both legs of the U-shape by displacement sensors, such as optical pickups. We have to analyze the movements in this device. The U-shaped tube is bent up and down. We interpret this as a rotary movement as indicated in the figure. Note: For a rotation vector, the vector arrow is perpendicular

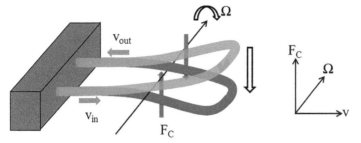

Figure 6.6 Coriolis flowmeter.
Look at the flow-in part of the tube: In the moment of the picture, v is positive (flow in), the rotation is positive (clockwise = down), and F_C is positive (up).
For the flow-out part of the tube, in the moment of the picture, v is negative (flow out), the rotation is positive (clockwise = down), and F_C is negative (down).

to the plane of rotation! The linear movement of the flow within the rotatory movement of the tube results in a Coriolis force. From the gyroscopes, we repeat the Coriolis force:

$$2m\vec{v} \times \vec{\Omega} = \vec{F}_C.$$

where m is the mass of the liquid in the vibrating tube.

The Coriolis force is perpendicular to the flow v and the rotation vector Ω, and thus it must point up or point down. Figure 6.6 shows the moment when the U-tube is moving down. In this moment, the rotation sense is mathematically positive. For the incoming flow, v is positive and the Coriolis force points upward. For the outgoing flow, the velocity v is negative and the Coriolis force points down. In sum, the two Coriolis forces will result in a twisting movement of the U-tube, as indicated in Figure 6.6. A moment later, the vibration of the U-tube is going up and then the Coriolis forces reverse direction, too. Figure 6.7 shows the impact on the vibration of the U-tube, seen from the front. Without flow, the movement is just up–down. With flow, a small twisting movement is superimposed. This is the secondary movement. What signal can we expect from the pickups of vertical movement? Without flow, they show a sinus vibration at same amplitude and same phase. With flow, we superimpose the small secondary movement. The signals will not be identical any more, with the main difference being a phase shift between the incoming and outgoing leg of the U-tube.

The page http://www.flowmeters.com/ explains several flow sensors with animations. An instructive animation of the movement of a Coriolis flowmeter can be found at http://en.wikipedia.org/wiki/Mass_flow_meter.

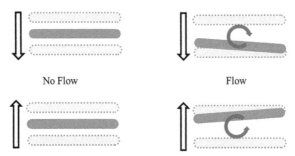

No Flow   Flow

Figure 6.7   Movement of the Coriolis flow sensor.
Left: Without the flow, the movement of the tube is just up–down (primary motion induced using an actuator).
Right: With flow, Coriolis force generates a superimposed twisting movement (secondary motion).

The Coriolis sensor obviously measures mass flow. But it can also measure density. The primary motion is a spring mass motion defined by the elasticity of the tube and by the mass of the tube plus the fluid in it. Measuring the resonance frequency, we can calculate the fluid density. Thus, a Coriolis flowmeter gives us values for both mass and volume flow.

The big advantage of the Coriolis flowmeter is that it can be made from a steel tube. It has no sensitive parts in contact with the flow. Accuracy is high, typically 0.1%. For aggressive media in industry, it is the first choice. It can also be used for difficult media such as pastes or two-phase flow, but then the accuracy may be less.

Classically, Coriolis flow measurement is a method for high mass flows. A steel system typically has from 1.500 to 600.000 kg/h full range, allowing temperature from −70 °C to +350 °C[1]. Accuracy is 0.1% for liquids and 0.5% for gases. Therefore, for liquids, 1.5 kg/h may be resolved. Recently, the company Bronkhorst has been demonstrating that Coriolis flowmeters can also be used for small flow rates[2] of liquids and gases. The smallest flow rate is 0.1–5 g/h of nitrogen gas, which corresponds to 1.3–67 ml/min. Resolution is 0.04 g/h. This device uses a folded U-tube that is made to vibrate by an electromagnetic actuator. Readout is done optically; two small mirrors are

---

[1]Yokogawa: http://www.yokogawa.com/de/fld/durchfluss/coriolis/index_coriolis/de-index
_coriolis.htm Endress+Hauser: https://www.de.endress.com/de/messgeraete-fuer-die-prozess technik/Durchflussmessung-Produkt%C3%BCbersicht/Coriolis-Proline-Promass-F300

[2]https://www.bronkhorst-nord.de/de/produkte/coriolis_durchflussmesser_und_-regler/

placed on the two arms of the U-tube, and the phase difference of their reflexes is measured.

Also, Coriolis flowmeters in MEMS technology have been shown[3]. Haneveld et al. made a sensor with a flow channel made of silicon nitride. Actuation is done by Lorenz force applying a current along the channel and an external magnetic field. Readout of the Coriolis-induced movement is done by comb electrodes. This device can measure liquids and gases and goes down to 1.2 g/h.

6.5 Electromagnetic Flow Sensor

Multiphase liquids such as pastes, ink, paint, sand in water, etc. are a special case. Due to their high viscosity, they are difficult to transport and measure. On the other hand, these liquids have a very interesting property: they carry ions. This means, when they flow, there is not only liquid flow but also an electric current $I = nq\dot{V}$, where n is the number of ions per volume, q their charge and $\dot{V}$ is the volume flow. This allows us to construct an electromagnetic flow sensor[4] (also called magnetic flowmeter or magneto-inductive flowmeter, or just magmeter) using Lorentz force. Can we measure water? Deionized water will not work, but tap water will. There are positive and negative ions, they are separated and we will find a voltage.

Essentially, the electromagnetic flow sensor shown in Figure 6.8 works like an inverted Hall sensor: We apply a magnetic field, measure the voltage generated and then determine the current, i.e., the number of ions coming by. This current is proportional to the mass flow. The liquid flows in an insulating plastic tube. There are two opposite electrode plates mounted in the tube and we apply a magnetic field. There will be a Lorentz force $\vec{F} = q\vec{v} \times \vec{B}$ on the charge carriers. They will concentrate on one side of the tube, as we have already discussed for the Hall sensor. We measure the charge accumulation using the electrode plates. The magnetic field is made by an electromagnet using AC excitation. Thus, lock-in technique can be used to reduce noise.

The advantage of electromagnetic flowmeters is that you can use it for high-viscosity materials such paste, mud, slurry or fluid containing particles. A second advantage is the large linear range. The pickup is voltage

---

[3]J. Haneveld, T. S. Lammerink, M. J. de Boer, R. G. P. Sanders, A. Mehendale, J. C. Lötters, M. Dijkstra, R. J. Wiegerink: Modelling, design, fabrication and characterization of a micro Coriolis mass flow sensor. J. Micromech. Microeng. 20 (2010).

[4]https://www.kobold.com/Magnetic-Inductive-Flowmeter-monitor-MIK

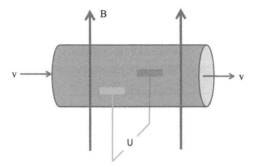

Figure 6.8 Electromagnetic flowmeter. It works like an "inverted Hall sensor". In a magnetic field, a flow is transporting ions. These are driven to one side by Lorentz force, which generates a voltage at the pickup electrodes.

measurement, which can be done over many orders of magnitude with high accuracy.

6.6 Thermal Flow Sensor

6.6.1 Hot Wire Probe

One of the classic flow sensors is the hot wire probe. It is a tungsten wire of 6-μm diameter and 3-mm length, which is soldered on a holder with two needles, as shown in Figure 6.9. We heat the wire by current and measure its temperature by its resistivity. The flow is cooling the wire, and thus the flow speed can be measured from the wire temperature.

A much better method is applying constant temperature at the wire. We construct a controller to keep the resistivity of the wire constant, and then also its temperature will be constant. Why is this better? Because it is faster! Let us compare a step response on increasing flow for constant current and constant temperature:

Constant current: More heat is taken off by the flow; the outer diameter becomes cooler. The thermal signal travels outside in; this takes some milliseconds. Finally, the whole wire is cooler.

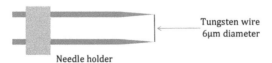

Needle holder

Tungsten wire
6μm diameter

Figure 6.9 Hot wire anemometer.

Constant temperature: As soon as the outer diameter starts getting cooler, the system increases heating. While the cooling impulse travels outside in, a heating impulse starts travelling inside out. Finally, both cancel out and the temperature is as before. This can happen in about 10 μs. The time constant is reduced to 10 μs, and we can achieve 100 kHz bandwidth! Constant temperature electronics for hot wire probes is the first method that was able to observe fast fluctuations in turbulent flow, and the understanding of turbulence made big progress by this. This is another nice example how active stabilization by a feedback loop makes the system faster, as we have seen for the accelerometer in Chapter 4, Figure 4.5.

6.6.2 The Mass Flow Controller

In many technical applications, you need a constant mass flow of a gas. An example is the supply of precursor gas to a plasma deposition machine. To obtain a repeatable film quality, the gas mass flow must be constant, independent of the pressures in the supply line and in the deposition chamber. This is done by a mass flow controller shown in Figure 6.10, which has a thermal mass flow sensor and a magnetic valve for actuation. Measurement is done in a bypass flow of the main mass flow. Around this tube, a thin wire is wound as an electric heater, which adds a specified amount of heat to the mass flow. The temperature upstream and downstream of the heater is measured with two more wire windings acting as resistive thermometers. They are connected in a Wheatstone bridge, and the control system keeps the bridge balanced acting on the magnetic valve. When, e.g., another machine is

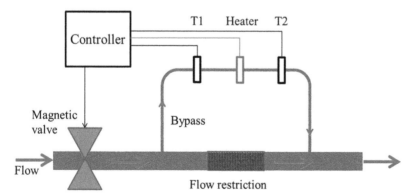

Figure 6.10 A mass flow controller. In the bypass flow, the flow is heated with constant power. The resulting temperature difference gives the flow. A magnetic valve is used to control for constant flow.

started and the pressure in the supply line drops, the valve will slightly open to regain the set point of mass flow.

This is an example for calorimetric measurement. There are two basic strategies for thermal flow measurement: We can put heat to the fluid and measure temperature increase (calorimeter). Or we can heat a body exposed to the flow and measure the power or temperature (anemometer). The advantage of calorimetry is the easy switching between different gases: when we have been calibrating to air, but we want to control an argon flow, we just need the values of specific heat of the two gases and we can calculate the correction factor.

6.6.3 Thermal Membrane Flow Sensors

Hot wire probes are powerful sensors, but difficult to handle, since the wire is easily broken. And they are expensive to make, since you have to weld each wire to the two needles. Why not use micromachining? This led to the thermal flow sensor[5]. The technology is well known; it is almost identical to the infrared thermopile. This has a thin membrane and thermometers, but it cannot be heated electrically. Hence, add a heating resistor to a thermopile and you obtain a thermal flow sensor, shown in Figures 6.11 and 6.12. There are two conducting layers for the thermopiles: polysilicon and tungsten-titanium (WTi). The heater in the middle is made in the WTi layer[6]. There are two rows of thermocouples upstream and downstream of the heater.

Figure 6.13 shows the basic idea: When there is no flow, there will be a hot cloud around the heater. It is symmetric and both thermopiles see the same temperature. When flow starts, there will be a flow of unheated fluid on the upstream thermopile, which is cooled. Then, the flow will be slightly heated by the heater and then go to the downstream thermopile. Figure 6.14 shows the thermoelectric voltage of the two thermopiles versus flow rate. The measurement is done with constant power at the thermopile. The upstream thermopile is increasingly cooled with increasing flow rate. The downstream

---

[5]Thermal Flow Sensor for Liquids and Gases based on combinations of two principles. M. Ashauer, H. Glosch, F. Hedrich, N. Hey, H. Sandmaier, W. Lang; Sensors & Actuators 73 (1999).

Wide range semiconductor flow sensors. A. Glaninger, A. Jachimowicz, F. Kohl, R. Chabicovsky, G. Urban. Sensors and Actuators 85 (2000).

[6]Buchner, R., C. Sosna, M. Maiwald, W. Benecke and W. Lang: A high-temperature thermopile fabrication process for thermal flow sensors. Sensors and Actuators A: Physical 130–131, 262–266 (2006).

Figure 6.11 Two-flow sensor chips on a 1 cent coin.

thermopile is also cooled, but less. The difference is caused by the thermal energy driven into the flow by the heater.

For measurement, we use the voltage difference given by the red curve in Figure 6.14. The curves are nonlinear. It is worthwhile comparing the thermal characteristics with the one of differential pressure flow sensing: both are essentially nonlinear, but the other way round. Thermal flowmeters are very sensitive at low flow rates; thus, they allow a high dynamic range. At high flow rates, the characteristic flattens, the sensors cannot resolve any more, but they are not destroyed. For differential pressure flow sensing, it was the other way round: flat characteristics at low flow rate and low dynamic range, plus the danger of membrane braking at overload.

Due to their high sensitivity at low flow rates, thermal anemometers are especially appropriate to measure very small flow rates. For air, a flow of 10 ml/h can be resolved[7]. Flow of water[8] can even be resolved down to 10 μl/h. For this experiment, the flow sensor is mounted in a small flow channel of 0.4 mm. 0.6 mm; this corresponds to a linear flow speed of

[7]Issa, S., W. Lang: Minimum Detectable Air Velocity by Thermal Flow Sensors. Sensors, Vol. 13(8), pp. 10944–10953 (2013).

[8]Thermal Flow Sensor for Very Small Flow Rate: M. Ashauer, H. Scholz, R. Briegel, H. Sandmaier, W. Lang. Transducers 2001, Munich, Germany, June 2001, Proceedings S. 1464–1467.

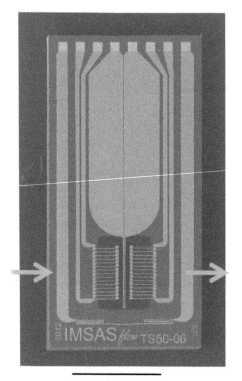

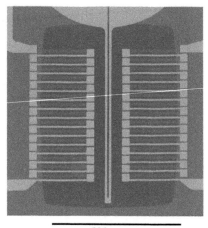

500 μm

1 mm

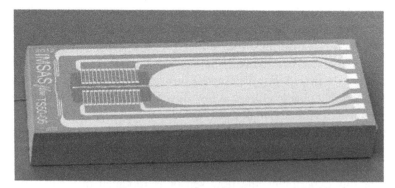

Figure 6.12 Thermal flow sensor on a nitride membrane as seen by the electron microscope. We see the heater in the middle, and one thermopile upstream and one downstream. The direction of flow is indicated by the blue arrows. Figures by R. Buchner, N. Hartgenbusch, E.-M. Meyer, IMSAS.

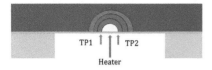
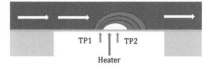

Without flow, the heater generates a symmetric heat cloud. The two thermopiles measure the same temperature.

With flow, the heat cloud is deformed and driven downstream. The upstream thermopile TP1 measures less temperature.

Figure 6.13   Heat distribution over a heated membrane.

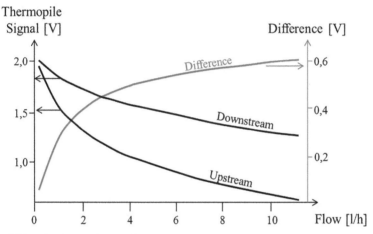

Figure 6.14   Membrane thermal flow sensor, constant power mode. Upstream temperature (TP1), downstream temperature (TP2) and difference signal. The thermal device is very sensitive for small flow rates; for high rates, it saturates.

11 μm/s. Even smaller flows down to 2.4 μl/h can be resolved when the flow channel is monolithically integrated on top of the sensor membrane[9,10] Concerning accuracy, Sensirion[11] (Figure 6.15) give 1.5% of the measured value for gases and 5% for liquids.

[9]Buchner, R., P. Bhargava, C. Sosna, W. Benecke, W. Lang: Thermoelectric Flow Sensors with Monolithically Integrated Channel Structures for Measurements of Very Small Flow Rates. Proceedings of IEEE Sensors 2007, pp. 828–831 (2007).

[10]S. Billat, K. Kliche, R. Gronmaier, P. Nommensen, J. Auber, F. Hedrich, R. Zengerle: Monolithic integration of micro-channel on disposable flow sensors for medical applications. Sensors and Actuators A 145–146 (2008).

[11]https://www.sensirion.com/fileadmin/user_upload/customers/sensirion/Dokumente/0_Datasheets/Mass_Flow_Meter/Sensirion_Mass_Flow_Meters_SFM3000_Datasheet.pdf.

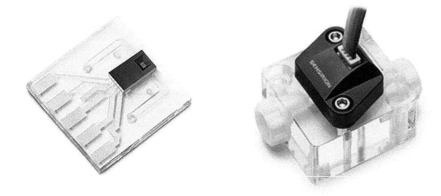

Figure 6.15   A commercial thermal flow sensor made via silicon technology using a thin membrane, type LPG10 by Sensirion[12].   Left: the sensor chip on a board. The chip is flipped; you can see the KOH etch grove and the membrane in it. Right: the sensor in a plastic housing. Figures by courtesy of Sensirion AG.

6.6.4 Modeling of the Thermal Anemometer

How can we understand the thermal flow sensor mathematically? We will calculate a model for the energy flow following the analysis of Lammerink[13].

We calculate a simple one-dimensional model; this can be imagined as a membrane sensor in a small flow channel, when thermal gradients across the chamber are neglected. Figure 6.16 shows the model assumptions. From the heater, thermal conduction transports energy upstream. From the inlet at negative x, there is a stream of cold fluid coming, which is a convective energy transport. Those two energy flows will determine a temperature profile

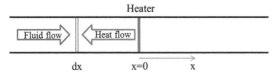

Figure 6.16   The one-dimensional model of a flow sensor. The heater is represented as a temperature boundary condition at $x = 0$. Heat flow from hot into cold competes with convective inflow of cold fluid.

---

[12]https://www.sensirion.com/de/durchflusssensoren/hochgenaue-fluessigkeitssensoren-fuer-kleine-flussraten/seite/lpg10-planare-miniatur-durchflusssensoren/

[13]T. S. Lammerink, N. R. Tas, M. Elwenspoek, J. H. Fluitman: Micro-Liquid Flow Sensor; Sensors and Actuators A, 37–38 (1993) 45–50.

Table 6.1   Input parameters for the thermal model of the flow sensor

		Water	Air
Thermal conductivity	K	0.6 W/m·K	0.024 W/m·K
Density	P	1 g/cm^3	0.001 g/cm^3
Thermal capacity	C	4.2 J/g·K	1 J/g·K

$T = T(x)$. The heater is represented as a temperature boundary condition at $x = 0$. The temperature of the incoming fluid is set to zero: $T(x \rightarrow -\infty) = 0$.

For <u>constant temperature</u> heating, the boundary condition is: $T(0) = T_0$;

We look at a small volume, a thin slice of the duct with thickness dx as seen in Figure 6.16.

For the energy flow, the general equation of heat flow is applied[14]:

$$Ak\frac{\partial^2 T}{\partial x^2} - A\rho cv\frac{\partial T}{\partial x} - q'' = A\rho c\frac{\partial T}{\partial t} \qquad (6.3)$$

$$\text{(I)} \qquad \text{(II)} \qquad \text{(III)} \qquad \text{(IV)}$$

where A is the cross section of the flow channel and k, ρ and c are the thermal conductivity, density and thermal capacity of the liquid, respectively. Values are given in Table 6.1.

(I) is the heat flowing from the heater into the inflowing cold fluid by heat conduction

(II) is the convective input of heat: the incoming flow transports cold fluid into the control volume. This is the inflowing heat capacity $A\rho cv$ multiplied with temperature gradient, which describes how much this incoming fluid has to be heated up.

(III) is the heat generation and loss. As the control volume does not include the heater, there is no generation. For the moment, we also will neglect heat loss from the fluid to the channel wall.

(IV) is the change of the temperature profile with time. Since we model a steady state, we can neglect this term.

Neglecting term (III) and (IV) Equation 6.3. reduces to

$$Ak\frac{\partial^2 T}{\partial x^2} = A\rho cv\frac{\partial T}{\partial x} \qquad (6.4)$$

or

$$\frac{k}{\rho cv}\frac{\partial^2 T}{\partial x^2} = \frac{\partial T}{\partial x} \qquad (6.5)$$

---

[14]D. Pitts, L. Sissom: Heat Transfer, Schaum's Outline Series 1997.

We assume the ansatz

$$T(x) = T_1 e^{\lambda x} \tag{6.6}$$

$$\frac{\partial T(x)}{\partial x} = \lambda T(x) \tag{6.7}$$

$$\frac{\partial^2 T(x)}{\partial x^2} = \lambda^2 T(x) \tag{6.8}$$

The characteristic equation is

$$\lambda^2 = \lambda \frac{\rho c v}{k} \tag{6.9}$$

There are two solutions:

$$\lambda_1 = \frac{\rho c v}{k} \quad \text{and} \quad \lambda_2 = 0 \tag{6.10}$$

The boundary condition $T(x = 0) = T_0$ results in $T_1 = T_0$. The temperature upstream the heater ($x < 0$) is

$$T(x) = T_0 e^{\frac{\rho c v}{k} x} \tag{6.11}$$

For the downstream part ($x > 0$), there is no heat exchange anymore; here, the second solution ($\lambda = 0$) applies and the temperature remains constant.

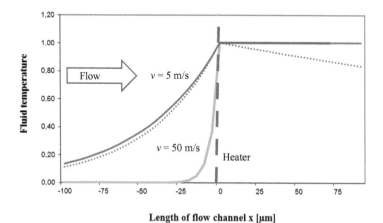

Figure 6.17   Modeled temperature profile over the membrane for airflow of 5 m/s and 50 m/s and for a constant temperature mode. When the flow is faster, the hot region is compressed close to the heater at $x = 0$. The dotted line is the model with an added loss term.

Figure 6.17 shows the result of the model; temperature is plotted versus length *x* for two flow speeds of air. For 5 m/s flow speed, we see that the temperature increases as expected when the fluid approaches the heater. This is done by the thermal conductivity of the fluid. After the heater, the temperature is constant, because we did not include a loss term yet. At a higher flow speed of 50 m/s, the figure looks similar, but the slope is compressed versus the heater.

This model helps us understand the most important design rule for thermal flow sensors: the critical parameter is the distance between the heater and the thermopiles. The exponent in Equation 6.11 is governed by the product of distance *x* and velocity *v*. When velocity is high, the distance must be small to achieve a similar characteristic curve. Hence, when we want to lay out for fast velocities, the thermopiles must be close to the heater. When we lay out for small velocities, the distance must be larger. For a range of 5 m/s, a distance of 50 μm would be appropriate. For 50 m/s, this sensor would show a flat characteristic; a distance of 10 μm would be much better then.

This can also be understood figuratively: Two heat flows are competing. The first is convective, which is proportional to velocity. The second is conductive, which is driven by a heat gradient. Fast flow will push the temperature profile close to the heater; then, the gradient increases and conductive transport is increased until it can cancel the flow. Therefore, we have to measure close to the heater for fast flow. It is interesting to note that this competing of convection and conduction is very similar to flame propagation: there also incoming gas is heated by the hot flame front by thermal conduction. A calculus similar to the design rule of flow sensors can be done to calculate the laminar flame speed in premixed gas flames[15].

Next we introduce a loss term. The flow will lose temperature when the fluid is warmer than the channel wall. To reflect this, we do not neglect the loss term q// in Equation 6.3. any longer. The loss will be proportional to the temperature difference. The characteristic equation will then be a quadratic equation. In Figure 6.17, the model with a loss term is added for a flow of 5 m/s (dotted line). We can see the fluid losing temperature when flowing away after the heater, but the response of the sensor is not changed fundamentally.

For <u>constant power</u> heating with power P_{in}, the boundary condition is a little more complicated. It can be determined from the power balance:

$$P_{in} = KT(0) + Ac\rho vT(0) \qquad (6.12)$$

---

[15]K. Kuo: Principles of Combustion, Wiley (1986), p. 293

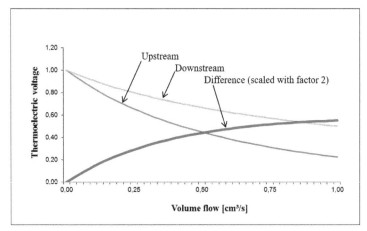

Figure 6.18   Thermal flow sensor in constant power mode modeling results for air. This corresponds to the measured characteristics in Figure 6.14.

where K gives the inevitable conductive loss of the heater, which happens also at zero flow; the second term is the heat needed to heat up the incoming fluid. Therefore,

$$T(0) = \frac{P_{in}}{K + A\rho cv} \qquad (6.13)$$

Figure 6.18 shows the modeled sensor response when constant power mode is applied; the thermoelectric voltage is plotted against flow velocity. As expected, we see that the temperature upstream falls with increasing flow. The temperature downstream also decreases, since a constant power mode is applied and the same power on more mass flow results in less heating. For sensing, the difference of the two thermoelectric voltages is calculated. It increases with temperature in a nonlinear way, as expected. When we exceed the measurement range (flow $>> 1$ cm³/s), all the heat generated will be drawn off by the flow. Upstream and downstream temperatures will approach zero and the difference too, of course. At very high flow rates, the sensor cannot measure any more, but it will not be destroyed.

Most thermal anemometers use constant power, because it is simpler to implement, but constant temperature is also applied. Recently, a new method of excitation has been discussed: pulse excitation[16]. The background is power

---

[16]Hartgenbusch, N.; Borysov, M.; Jedermann, R.; Lang, W., *Reduction of power consumption and expansion of the measurement range by pulsed excitation of thermal flow sensors.* Sensors and Actuators A: Physical, 2017.

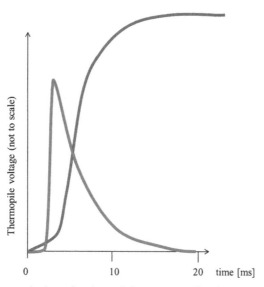

Figure 6.19 Pulse excitation of a thermal flow sensor. Classic method of waiting for a stationary state compared to the new method using a short pulse.

consumption. When we want to use wireless sensor nodes, save battery and use a duty cycle for the sensors, most of the time, they should be in sleep mode. A flow sensor has a thermal time constant of about $\tau = 5$ ms. In order to obtain a steady state measurement, I should wait $5 \cdot \tau = 25$ ms for the signal to stabilize, as shown in Figure 6.19. But is it really necessary to wait so long? The alternative is pulse excitation: I heat with a short pulse and then measure the response pulse of the sensor, without ever waiting for a steady state. Using this method, the average power consumption could be reduced from 1 mW to 100 μW. But also, when there is no need to save power, pulse excitation has an advantage: the dynamic range is increased, i.e., we can apply a higher flow rate before the response curve flattens.

Questions

- Is the volume flow a scalar or a vector?
- What types of flowmeters do you know?
- What is a Venturi tube? Write down and explain Bernoulli's equation. How is it implemented practically? Explain the characteristics. What problems does nonlinearity cause?
- Explain the Coriolis flowmeter. What are its advantages?

- Explain the electrodynamic flowmeter. What is it used for?
- What is a hot wire probe?
- Why is constant temperature heating of hot wire probes better than constant voltage heating?
- Draw a thermal flow sensor
- Sketch the voltage output of a thermal flow sensor in constant power mode

Index

About the Author

Walter Lang studied physics at Munich University and received his Diploma in 1982 on Raman spectroscopy of crystals with low symmetry. His Ph.D. in engineering at Munich Technical University was on flame-induced vibrations. In 1987 he joined the Fraunhofer Institute for Solid State Technology in Munich, where he worked on microsystems technology. In 1995 he became the head of the sensors department at the Institute of Micromachining and Information Technology of the Hahn-Schickard Gesellschaft (HSG-IMIT) in Villingen-Schwenningen, Germany, working on sensors for flow and angular rate, sensor test and modelling. February 2003 Walter Lang joined the University of Bremen. He is heading the Institute for microsensors, -actuators and –systems (IMSAS) and he is the speaker of the Microsystems Center Bremen (MCB). His projects cover thermal sensors, sensor networks for logistics and the embedding of sensors in materials.

Publications: http://scholar.google.de/citations?user=hveaqzoAAAAJ&hl=de